MACHINE LEARNING IN ASTRONOMY: POSSIBILITIES AND PITFALLS

IAU SYMPOSIUM 368

COVER ILLUSTRATION:

'Word cloud generated from symposium abstracts'
(C) Jess McIver, Ashish Mahabal, Christopher Fluke

IAU SYMPOSIUM PROCEEDINGS SERIES

INTERNATIONAL ASTRONOMICAL UNION

UNION ASTRONOMIQUE INTERNATIONALE

International Astronomical Union

IAU

MACHINE LEARNING IN ASTRONOMY: POSSIBILITIES AND PITFALLS

PROCEEDINGS OF THE 368th SYMPOSIUM OF THE INTERNATIONAL ASTRONOMICAL UNION BUSAN, REPUBLIC OF KOREA 2–4 AUGUST, 2022

Edited by

Jess McIver
University of British Columbia, Canada

Ashish Mahabal
California Institute of Technology, USA

and

Christopher Fluke
Swinburne University of Technology, Australia

CAMBRIDGE
UNIVERSITY PRESS

CAMBRIDGE UNIVERSITY PRESS
University Printing House, Cambridge CB2 8BS, United Kingdom
1 Liberty Plaza, Floor 20, New York, NY 10006, USA
10 Stamford Road, Oakleigh, Melbourne 3166, Australia

First published 2025

Printed in Great Britain by Henry Ling Limited, The Dorset Press, Dorchester, DT1 1HQ

Typeset in System LaTeX 2ε

A catalogue record for this book is available from the British Library Library of Congress Cataloguing in Publication data

This journal issue has been printed on FSC™-certified paper and cover board. FSC is an independent, non-governmental, not-for-profit organization established to promote the responsible management of the world's forests. Please see www.fsc.org for information.

ISBN 9781009345194 hardback
ISSN 1743-9213

Table of Contents

Preface

Symposium 368 of the IAU GA in Busan on "Machine Learning - Possibilities and Pitfalls" was held from Aug 2 to 4 2022. The theme revolved around machine learning in astronomy, in particular the potential it has created to solve many open challenges, but also the possible problems that can be encountered when applying these often black-boxy techniques without proper care.

In the symposium there were eight invited and 13 contributed talks. There were also four panel discussions during which 15 total panelists shared the stage. For want of time another $\sim$70 abstracts were converted to virtual talks and posters available throughout the meeting.

The symposium started with two detailed tutorials: "ML tutorial for the broader community" and "Classic Machine Learning vs Deep Learning: when, why and how?". These provided a gentle introduction to those who were not steeped in the details of machine learning (ML), giving them the impetus to be able to follow the other talks later. For those already well-versed they served as a quick revision.

The plenary talks addressed the trends and challenges in deep learning, addressing how deep learning contrasts with shallow learning, the challenges of explainability and interpretability as more deep learning techniques are embraced, and the path to collaborative Human-AI learning addressing how with modern ML/AI methods for the first time we are starting to discover patterns from data-driven approaches and how this leads to areas not investigated before.

Other invited talks included reviews of various areas, including an overview of machine learning in astronomy, existing and forthcoming large and diverse datasets that form staple inputs to ML algorithms, the role of physics-informed machine learning in exoplanet characterization, and citizen science aspects for creating labeled datasets and for outreach and education.

The discussion sessions covered various aspects of the advances in ML, and the possible unintended misuse thereof. The first one discussed the breadth of ML applications in astronomy, highlighting the fact that for many problems classical ML techniques may be the first refuge, but also that recent improvements in techniques mean that one does not need huge datasets to do effective machine learning. Another discussed the important topic about combining diverse datasets, in particular archival datasets with real-time data. A third took on practical problem solving, including interpretability. Finally, the fourth discussion handled more recent trends in multi-messenger astronomy. These open discussions saw strong participation by students and postdocs.

The contributed talks - chosen from a large pool of abstracts - provided more extensive coverage of the field, touching upon more surveys, outliers and anomalies, unsupervised methods, incorporation of errors, non-Gaussianity, Generative Adversarial Networks, etc. The speakers included not only senior researchers but also undergraduate and graduate students. Many of the topics the invited and contributed talks did not cover were covered by the virtual talks and posters, providing a well-rounded symposium with science topics ranging from our neighborhood to cosmology.

The symposium received overwhelming support with close to 450 astronomers wanting to attend the symposium in person, and about 100 submitting abstracts for presentations. We tried to select representative presentations from across the spectrum, with the talks covering theory, simulations, applications, interpretations and on datasets from the Solar System to Galactic to extragalactic astronomy to cosmology. There were many deserving abstracts that had to be converted to virtual talks and posters.

We ensured that there was ample time for discussion through various panels on topics like Gravitational Wave/Multi-Messenger Astronomy, broader ML, and fusion of large datasets, in which all attendees were able to participate. For example, he challenges that ML is facing include the lack of interpretability and explainability. On another level, not many techniques allow for proper uncertainty quantification. While we did not expect the symposium to solve these big problems, we did manage to grow awareness amongst a larger set of users and practitioners of the issues, moving the needle towards better ML practices.

The broad hope and expectation was that attendees gain exposure to the breadth of available datasets and techniques, and the expertise on display during the symposium allows them to step out of their comfort zone to take on bigger problems while leveraging best practices. In particular, we hope to see more population studies incorporating more publicly available datasets (including transfer learning across datasets) rather than specialized studies involving smaller private and proprietary datasets, and we believe the symposium has helped the field move in that direction.

Since the pandemic had not yet abated, and the corresponding health safety measures were still in place, we likely lost many interested attendees (even from the 15 SOC members, only five could attend in person). Also, there were many sessions in parallel further dividing the attendees. Yet, throughout the symposium over 100 attendees were present plus many online (the exact number was not reported). There were a few complaints from those online about not being able to follow everything in the room - this is the unfortunate reality of hybrid meetings.

There were ample opportunities for discussions which the attendees took advantage of. There was dedicated time for poster viewing; since these were virtual posters it is likely that they got less visibility during the meeting from those in person, however, having the posters available to view online for an extended period may have resulted in more overall visibility in the field.

Overall the symposium set things up well to invite novices into the world of machine learning with datasets large and small, and provided the experts with more fodder to explore new problems, datasets, and techniques while providing all attendees with the resource to take steps towards avoiding the associated pitfalls.

Editors

Jess McIver
University of British Columbia, Canada

Ashish Mahabal
California Institute of Technology, USA

Christopher Fluke
Swinburne University of Technology, Australia

Scientific Organising Committee

Local Organising Committee

List of Participants

Aisha Alowais
Alisher S. Hojaev
Amelia Yu
Annalisa Pillepich
Anupam Bhardwaj
Arman Shafieloo
Ashish Mahabal
Atila Poro
Atrideb Chatterjee
Bekdaulet Shukirgaliyev
Boon Kiat Oh
Connor Bottrell
Cristiano Sabiu
Da Eun Kang
David Parkinson
Dennis Crake
Didier Fraix-Burnet
Djordje Savic
Dominic Adams
Dongjin Lee
Eric Ford
Erik Rodrigues de Lima
Eunsu Park
Garreth Martin
George Djorgovski
Georgios Vernardos
Gor Mikayelyan
Gordian Edenhofer
Guillermo Torralba Elipe
Guillermo Cabrera
Hanwool Koo
Ignacio Ferreras
Jaehan Bae
Jakob Knollmüller
Jeroen Audenaert
Jinkoo Yim
Joeri Van Leeuwen
Johan Knapen
John Suárez-Pérez
Josef Durech
Joshua Speagle
Juan Rafael Martinez Galarza
Kai Polsterer
Krisztián Vida
Kyungmin Kim
Laurent Eyer
Lilianne Nakazono
Lukasz Wyrzykowski
Marc Huertas-Company
Marco Álvarez
Margherita Grespan
Maria Luisa Buzzo
Marian Douspis
Mark Birkinshaw
Médéric Boquien
Melissa Lopez
Michel Cure
Michelle Lochner
Mike Walmsley
Monika Soraisam
Nadejda Blagorodnova Mujortova
Nahathai Tanakul
Nicolás Vásquez
O. Ivy Wong
Ofer Lahav
Oliver Boersma
Pablo Manuel Sánchez Alarcón
Priya Shah
Rania Hamdani
Raquel Ruiz Valença
Renee Hlozek
Rodrigo Calderon
Ryan Keeley
Ryohtaroh Ishikawa
Syed Najamul Hasan
Sandor Pinter
Sara Webb
Seong-Heon Lee
Sepideh Ghaziasgar
Shay Zucker
Simon Anghel
Sirinrat Sithajan
Stefan Wagner
Steffani Grondin
Suhyun Shin
Suk Yee Yong
Sumi Kim
Sungwook E Hong
Tomasz Różański
Vanessa McBride
Vishal Upendran
Vitor Cernic
William Pearson
Wuhyun Sohn
Yuan-Sen Ting

Machine Learning in Astronomy: Possibilities and Pitfalls
Proceedings IAU Symposium No. 368, 2025
J. McIver, A. Mahabal & C. Fluke, eds.
doi:10.1017/S174392132300128X

Enhancing Exoplanet Surveys via Physics-informed Machine Learning

Eric B. Ford[1,2,3,4]

[1]Department of Astronomy & Astrophysics, 525 Davey Laboratory, The Pennsylvania State University, University Park, PA, 16802, USA
email: ebf11@psu.edu

[2]Institute for Computational and Data Sciences, The Pennsylvania State University, University Park, PA, 16802, USA

[3]Center for Astrostatistics, 525 Davey Laboratory, The Pennsylvania State University, University Park, PA, 16802, USA

[4]Center for Exoplanets and Habitable Worlds, 525 Davey Laboratory, The Pennsylvania State University, University Park, PA, 16802, USA

Abstract. Since the last 1990s, Doppler spectroscopy has been one of the most prolific methods of detecting and characterizing exoplanets (Fischer et al. 2016). The latest generation of stabilized spectrographs can achieve impressive levels of precision and stability, approaching that needed to detect the motion of a Sun-like star due to the gravity of an Earth-mass planet in its habitable zone (Crass et al. 2021). However, the exoplanet detection power of modern radial velocity (RV) exoplanet surveys is typically limited by the spectral variability of the target star. Machine learning (ML) has the potential to significantly improve the ability of RV exoplanet surveys to distinguish planets for stellar variability. Astronomers have begun making applying a wide variety of ML techniques, from principal component analysis and multilinear regression to convolutional neural networks. This paper reviews the state of the field for mitigating stellar variability in RV exoplanet surveys from a ML perspective. Early results show that relatively simple ML techniques paired with well-engineered features often perform comparable to much more complex ML models, while providing improved interpretability and explainability. These are likely to be critical factors for establishing the credibility and robustness of any future detections of potentially Earth-like planets.

Keywords. methods: data analysis – methods: statistical – planets and satellites: detection – techniques: radial velocities, spectroscopic

1. Introduction

The problem of spectral variability is not new or unique to the search for Earth-mass planets (e.g., Queloz et al. 2001). In the early days of RV exoplanet surveys relied on relatively simple mitigation strategies. For example, one could search for exoplanet candidates based on an RV time-series, and then attempt to reject the null hypothesis that there were no variations in stellar activity indicators or spectral line shapes with the same periodicity as the putative planet. While this approach was effective in recognizing many potential false positive giant planets, it has limited utility for Earth-mass planets, since it is much harder to recognize the corresponding changes in activity indicators and line shapes due to the small-scale active regions that can mimic the Doppler signature of low-mass planets (Haywood et al. 2016).

Another common strategy was simply to avoid stars with significant variability by focusing on "RV quiet" stars. However, as astronomers have developed more precise instruments, we've discovered that most stars show intrinsic spectral variations capable of dwarfing the signature of Earth-mass planets. Further, future direct imaging missions will demand that RV surveys characterize the masses of planets around stars sufficiently nearby to permit direct detection (Crass et al. 2021).

As the instrumental precision of exoplanet-hunting spectrographs has improved, astronomers have considered a variety of statistical approaches to detect low-mass planets with varying levels of statistical rigor and success. Some challenging discoveries have stood the test of time (e.g., Haywood et al. 2014), while others have been refuted (e.g., Rajpaul et al. 2016). Unfortunately, there is not a practical first principles model for the effects of stellar variability on the spectrum. On the plus side, this presents an opportunity for scientists with expertise in statistics and ML to enhance the capabilities of ongoing and future exoplanet searches.

The RV signature of exoplanets has two key characteristics, it imprints on the stellar spectrum as a pure Doppler shift at each epoch, and the pattern of Doppler shifts in time is precisely described by the laws of Newtonian gravity. Intrinsic stellar variability causes changes in the spectroscopic timeseries that do not respect these constraints. In principle, one should be able to distinguish planets from stellar variability, given sufficient quantities of high-quality data. However, stellar variability can mimic the spectral signature of a planet and may be indistinguishable for a given data set. Astronomers have yet to converge on a statistical model or algorithm for distinguishing exoplanets from intrinsic stellar variability.

This paper provides an overview of recent progress toward mitigating stellar variability and a perspective on promising areas for future research in how ML can help RV exoplanet surveys realize their full scientific potential. §2 focuses on using information in individual spectra, while §3 focuses on using information in the temporal domain. For a more mathematical review of the state-of-the-art in measuring RVs, as well as a more complete set of references, see the recent review by Hara & Ford (2023). §4 discusses recent steps towards combining these into a unified framework. The paper concludes by offering perspectives on the future of ML and RV planet searches in §5.

2. Stellar Variability in the Wavelength Domain

2.1. *Template Fitting & Forward Modeling*

First, we consider attempts to improve the accuracy of Doppler planet surveys based on using the information in the wavelength domain. If one assumes a constant stellar spectrum, then there are multiple approaches to estimating the star's RV. The most basic straightforward approach to measure a Doppler shift between similar spectra is to model each spectrum as a shifted version of a common "template" spectrum (Zechmeister et al. 2018, Silva et al. 2022). For small Doppler shifts, the shift can be efficiently computed by projecting the residuals between an observation and the template onto the derivative of the template spectrum. The template fitting approach works particularly well for regions of the spectrum with many blended lines or broad molecular features where there are few well-isolated spectral lines. The template spectrum could be estimated by simply averaging spectra (after interpolating onto a common set of wavelengths in the solar system barycentric frame) or by invoking ML techniques such as Gaussian process deconvolution (to account for line spread function and perhaps other signals such as scattered light or telluric absorption).

A simple PCA analysis of modern spectrographs and exoplanet surveys is capable of recognizing the effects of magnetic activity on the stellar spectrum (Davis et al. 2017).

While the standard template fitting technique does not account for stellar or telluric variability, it can be generalized into a forward-modeling approach that allows for telluric variability (Bedell et al. 2019), variability in the stellar spectra (Jones et al. 2022) or temporal variability in both the stellar spectra and telluric absorption (Gilbertson et al. 2024). These physics-informed ML methods combine the benefits of being very flexible and strictly respecting physical constraints (e.g., Doppler shift, telluric model only removing flux, fluxes begin non-negative). Physics-informed ML methods can offer improved sensitivity since they do not need to "learn" the physics that has been into the model. Further, they result in model parameters (e.g., basis vectors) that are more readily interpretable than more complex ML methods such as neural networks.

As astronomers have begun thinking in terms of physics-informed ML models for high-resolution stellar spectroscopic time-series (e.g., Gully-Santiago & Morley 2022; Gilbertson et al. 2023), they can "rediscover" some of the classical activity indicators (e.g., Ca II H & K emission). One on hand, these may appear to be "obvious" to traditional astronomers. On the other hand, demonstrating the ability of methods to recover well-known results within a specific formalism can provide a valuable means of establishing the credibility of sophisticated algorithms and complex statistical models. In some cases, they may even provide new insights into the classical results.

2.2. *Dimensional Reduction via Cross-correlation Functions*

Most of the Doppler information is concentrated in a small fraction of the spectrum very near stellar absorption lines. Therefore, astronomers commonly estimate the stellar RV based on the cross-correlation function (CCF) of the stellar spectrum with either a stellar template spectrum or a "mask" with non-zero entries near spectral lines (Bouchy et al. 2001). The CCF can be interpreted as: 1) computing an "average" spectral line shape, 2) a feature engineered to serve as an input for ML algorithms to learn about line shape variations, and/or 3) a form of physics-informed dimensional reduction. By combining hundreds or thousands of spectral lines, the CCF can be characterized at much higher signal-to-noise (SNR) than individual spectral lines. This allows for measuring precise RVs (estimated by either fitting a Gaussian to the CCF, or comparing the CCF to a template CCF as above for the original spectrum), as well as CCF shape diagnostics such as the CCF bisector (mid-velocity for each CCF depth) and summary statistics related to the CCF bisector (e.g., velocity span of the CCF bisector at two depths or slope of the bisector between certain depths). Choosing CCF shape diagnostics can be interpreted as another form of feature engineering and dimensional reduction. While these can be useful for recognizing datasets where stellar variability could be responsible for a putative planetary RV signal, in some cases the information lost by such a dramatic dimensional reduction can lead to stellar variability being mistaken as a planet (e.g., Queloz et al. 2001). One opportunity for ML is to help select the most effective of the traditional summary statistics, to refine them (e.g., optimizing the choice of depths at which the CCF bisector span is measured) or to build new ones (e.g., functions of the CCF bisector).

Recently, multiple research groups have applied ML techniques to the CCF in an effort to learn how spurious apparent RVs can be recognized from CCF shape changes. de Beurs et al. (2022) demonstrated that applying multi-linear regression, a dense neural network or a convolutional neural network to the CCF can significantly reduce the effects of solar activity on inferred RVs. The FIESTA 2 algorithm (Zhao et al. 2022) achieved similar results with multi-linear regression but provided improved interpretability by first projecting the CCF onto a Fourier basis. They showed that most information useful for correcting spurious apparent RVs is contained in just the first several components. The SCALPELS algorithm (Collier Cameron et al. 2021) also achieved comparable results applying a series of transformations to the CCF (i.e., autocorrelation and SVD decomposition) prior to applying multilinear regression. The time-series of coefficients for several components were clearly linked with specific physical effects (e.g., CCF width variations due to Earth's eccentricity, instrumental interventions), while the coefficients for other components appear to be dominated by intrinsic stellar variability (based on their characteristic timescales).

Traditionally, astronomers have worked from a single CCF per spectrum. In practice, astronomers usually compute multiple CCFs based on small portions of the spectrum and then combined them to reach a single CCF for the entire spectrum. However, recent work has begun to explore using information from multiple CCFs per spectrum (e.g., computed for different wavelength ranges (Zechmeister et al. 2019, Cale et al. 2021) or lines likely to be more/less sensitive to stellar variability (Wise et al. 2022). This appears to be a promising avenue for ML to explore further.

2.3. *Line-by-line modeling*

Recently, astronomers have begun modeling the changes in the spectrum near each of hundreds or thousands of spectral lines separately (Dumusque et al. 2018, Cretignier et al. 2020, 2021; Artigau et al. 2022). The precision of the RVs estimated from each line is much lower than the signal of giant planets, and line shape variations can not be recognized for most individual spectral lines. For a pure Doppler shift, all lines shift by the same velocity, and an appropriately weighted average of the line-by-line RVs should achieve a precision comparable to measuring the Doppler shift of the CCF computed from the same lines. While intrinsic stellar variability can still cause a spurious apparent Doppler shift when RVs are estimated this way, the line-by-line approach offers the potential advantage of providing more information to recognize signatures of stellar activity (e.g., estimates of each line's depth, width, and/or shape). Measuring individual line properties from each spectrum can be interpreted as an alternative form of feature engineering and dimensional reduction that is less extreme than the CCF and preserves information about the different responses of different lines to stellar variability.

Studying the properties of spectral lines that have large (small) correlations with the bulk velocity provides insight into why many lines are sensitive (insensitive) to stellar variability (Dumusque et al. 2018, Wise et al. 2018). For example, many of the lines that initially appear to be most activity sensitive are blended with nearby lines. Simply selecting lines that avoid significant blending can reduce the magnitude of spurious RVs. However, aggressively avoiding blended lines can dramatically reduce the number of lines available and the resulting Doppler precision. Both theoretical models and empirical analyses suggest significant trends in apparent RV with line depth (Wise et al. 2022) and/or formation temperature (Al Moulla et al. 2022). This motivates further feature engineering (e.g., computing CCFs with curated line lists) or even new physics-informed ML algorithms (e.g., Cretignier et al. 2022).

3. Stellar Variability in the Time Domain

Next, we consider attempts to improve the accuracy of Doppler planet surveys based on using the information in the time domain. The RV signature of a low-mass planet has a constant amplitude, phase, and shape. The RV signature of each planet is typically very well described by Keplerian orbit, and can often be approximated as sinusoidal for low-mass planets. (While multiple-planet systems can cause detectable deviations from a superposition of Keplerian orbits, these are typically undetectably small over the relatively short timespan of exoplanet surveys for systems with low-mass planets.) While the spurious RV signals due to stellar variability can appear similar to a planet, Bayesian model comparison can be used to evaluate whether the observations are more consistent with a planet or stellar variability.

The most common method for jointly analyzing stellar variability and putative planetary signals in the time-domain is to model the observed radial velocity as the sum of Doppler shifts due to planetary motion, spurious signals due to stellar variability, and measurement noise. The measurement noise is often modeled as uncorrelated, Gaussian noise (after applying calibration corrections to remove temporally correlated noise). A similar white noise model for stellar variability can be effective for RV surveys with few closely spaced observations (Ford 2006). However, modern RV surveys targeting low-mass planets often make multiple observations on consecutive nights or even within one night. For these surveys, alternative strategies are necessary.

There are several physical mechanisms that cause stellar spectral variability. Unfortunately, there is not a practical first principles model for performing inference on a complete model for stellar pulsations, granulation or magnetic activity. Nevertheless, astrophysical knowledge of the timescales for each mechanism motivates observational strategies that mitigate the impact of that mechanism on observed RVs and informs modeling choices once data has been collected.

3.1. *Observational Strategies*

Some forms of stellar variability can be effectively mitigated by integrating the star light over a longer time span. This works remarkably well for mitigating p-mode pulsations since they have timescales of minutes (Dumusque et al. 2011, Chaplin et al. 2019). However, it is less effective in mitigating the effects of granulation (Meunier et al. 2015, Cegla et al. 2019, Luhn et al. 2023). Supergranulation is a particular concern, as it can induce correlations in observations spanning multiple nights.

Other mechanisms of stellar variability (e.g., rotation and evolution of active regions, magnetic activity cycles) have much longer timescales. In the idealized case that stellar variability presents as white "noise" in the RV time-series, the detection sensitivity would improve as $\sim 1/\sqrt{N_{\mathrm{obs}}}$, so simply collecting many observations would be a viable strategy (Newman et al. 2023). However, if one accounts for the temporal correlations of stellar variability, then the sensitivity improves more slowly and a very large number of observations are needed (Gilbertson et al. 2020a, Nicholson et al. 2022, Luhn et al. 2023). This motivates astronomers to upgrade from treating stellar variability as a noise to treating it as a signal that can be modeled jointly with the RV time-series.

3.2. *Gaussian Processes*

Gaussian processes (GPs) are a powerful tool for modeling complex physical processes, such as the effects of stellar activity, as they can incorporate physical constraints/intuition (via choice of the GP mean or covariance kernel) and permit the propagation of uncertainties associated with stellar variability into the posterior distribution for planetary parameters and the evidence (a technical term for the marginalized likelihood) used for Bayesian model comparison. In principle, one can easily replace a white noise model for stellar variability with a GP. For some mechanisms, the physics motivates a specific kernel choice (e.g., stochastically excited harmonic oscillator for pulsations; Brewer & Stello 2009, Luhn et al. 2023). For large data sets, computational considerations may dictate using a different, but qualitatively similar kernel (Foreman-Mackey et al. 2017; Delisle et al. 2022). For stars with substantial rotationally-modulated stellar variability (e.g., spots, facula), there is not a "correct" kernel, but physical considerations inform the kernel choice. The quasi-periodic (QP) kernel is a common choice for stars with large rotationally modulated variability (Aigrain et al. 2012). For stars with lower levels of activity, the typical spot can appear and decay in less than one rotation period, in which case Matern 5/2 kernels outperform the QP kernel (Gilbertson et al. 2020a). While GPs are clearly useful tools for mathematical modeling, there is a risk of overinterpreting their parameter values as corresponding to physical parameters (e.g., rotation rate, spot decay time), rather than parameters in a mathematical model (i.e., the Greek letter fallacy). It is particularly important to keep this in mind when comparing models that share parameter names in common.

Thoughtful choice of priors for GP kernel parameters is important to minimize bias in planetary parameters. Regardless of the choice of kernel, the mean of a GP (with zero mean prior) conditioned on observations at any point is just the weighted sum of observations. If the GP kernel's characteristic timescale is short, then a one-dimensional GP's posterior mean can track the observations, resulting in residuals consistent with measurement uncertainties and potentially absorbing a planet as well as the stellar variability. Even if the most extreme examples of this behavior can be avoided by placing a prior on the GP length scale, including a GP term when modeling a dataset of just RVs will still be prone to inducing a bias in the RV amplitude (and thus planet mass).

When both RVs and complementary observations (e.g., photometry, activity indicators) are available, it is preferable to model both types of observations with a shared latent GP. The $F - F'$ model for photometric and RV variability is an example of developing a physics-informed multivariate GP model (Aigrain et al. 2012). Unfortunately, contemporaneous high-precision photometry is often not available for stars targeted by RV surveys. Several generalizations of the $F - F'$ that use RVs and some form of activity indicators have proven useful for modeling RV survey observations (Rajpaul et al. 2015; Gilbertson et al. 2020a, Barragán et al. 2022, Delisle et al. 2022, Jones et al. 2022).

When a latent GP is conditioned on multiple types of contemporaneous observations, the residuals relative to the GP's posterior mean are likely to be significantly larger than the measurement uncertainties for realistic stellar variability (Rajpaul et al. 2015; Aigrian & Foreman-Mackey 2022). This allows for a quantitative assessment of the extent of model misspecification due to the use of a GP to model complex stellar astrophysics.

While most previous work has used stationary GP kernels, the extent of stellar variability can vary substantially over a stellar activity cycle. In principle, a GP using a non-stationary kernel or a generalization such as a Gaussian process regression network (Camacho et al. 2023) appear promising avenues for further research.

4. Putting the Pieces Together

There is still room for improvement in modeling either the wavelength and temporal variability separately. Eventually, these should be integrated into a single model that simultaneously models both signatures of stellar variability, so as to maximize astronomers' ability to distinguish exoplanets and stellar astrophysics. In principle, either traditional activity indicators or data-driven stellar variability indicators can be incorporated into multi-variate GP models (e.g., Gilbertson et al. 2020a; Zhao et al. 2022). In practice, astronomers will need to experiment with various choices for variability indicators and GP kernels and build intuition for how to perform inference effectively with such complex models.

Given the existence of a variety of strategies for mitigating stellar variability, it is natural to want to compare them to identify which are "best" for exoplanet hunting. In a recent community data challenge, several research groups applied 22 methods to analyze EXPRES observations of RV quiet stars (Zhao et al. 2020). Nearly every method applied appeared to reduce spurious RV signals due to stellar variability, when considered individually. Of course, it is to be expected that there would be a difference in the predicted spurious RV signal, if some methods were more effective than others. However, a more detailed comparison revealed that different methods did not agree on how much of the measured RV was due to a spurious signal from stellar variability (or even agree on the sign of the spurious perturbation to the RVs; Zhao et al. 2022). Since the true RV of the target stars is unknown, it was impossible to know which (if any) methods actually resulted in improved RVs. This presents a significant challenge for training, testing and verifying ML algorithms designed to mitigate stellar variaibilty.

5. Future

The combination of state-of-the-art instruments, large allocations of observing time and modern ML and emerging physics-informed ML techniques have the potential to enable the detection of Earth-mass planets around Sun-like stars. However, the importance of data analysis, statistical modeling and perhaps ML can not be underestimated. Characterizing Earth-mass planets in the habitable zone on Sun-like stars with RVs will depend on the success of ongoing research to mitigate stellar variability. Simulations of large, next-generation exoplanet surveys suggest that detecting and characterizing an Earth-analog with RVs will likely require over a decade of intensive observations, largely due to stellar variability (Nicholson et al. 2022, Luhn et al. 2023). Multiple research groups are exploring multiple strategies and making technical improvements. Even modest improvements in the data analysis and inference pipeline are likely to have a significant impact on the number and timespan of observations (and hence the cost) required to characterize potential Earth-analogs.

However, some fundamental challenges have yet to be satisfactorily resolved. First, how can astronomers validate sophisticated ML and statistical models for spectroscopic time-series? Second, even if a small number of astronomers fluent in Bayesian inference and modern ML techniques convince themselves that they have discovered and characterized an Earth-analog planet, how could they convince other astronomers that their favored combination of ML-based methods and sophisticated statistical modeling is trustworthy and robust?

One approach is to train and validate ML and statistical methods on synthetic datasets. First principles, physics-based simulations are too computationally demanding (e.g., Dravins et al. 2021). An intermediate strategy is to synthesize synthetic spectra making use of spatially resolved observations of the sun. The community has developed multiple codes for generating synthetic spectra that include effects of active regions (Dumusque et al. 2014; Gilbertson et al. 2020b; Zhao et al. 2023) and granulation (Meunier et al. 2010; Palumbo et al. 2022). They offer the advantage of being able to synthesize large datasets and to perform experiments to see how the results respond to specific changes (e.g., changing the rotation period or viewing geometry). However, existing spectral synthesis codes provide a significantly simplified approximation of the full effects of stellar variability. For example, data synthesized by SOAP 2.0 can be very accurately modeled as the sum of a template spectrum, Doppler shift, and one or two basis vectors (Davis et al. 2017, Jones et al. 2022; Zhao et al. 2022), while actual observations of the sun require at least several feature vectors (Collier Cameron et al. 2019; Zhao et al. 2022). Therefore, it is not surprising that while many methods are very effective at mitigating simulated stellar variability, they typically do not perform as well when tested on actual observations (Collier Cameron et al. 2019; de Beurs et al. 2022; Zhao et al. 2022; Perger et al. 2023).

There is one star for which the true velocity is precisely known, the Sun. Several exoplanet hunting spectrographs now routinely observe the Sun during the day, so as to help understand any instrumental variations and to help develop datasets for training future ML-based methods to mitigate stellar variability (Dumusque et al. 2015, 2021; Collier Cameron et al. 2019; Lin et al. 2022). One advantage of training on Sun-as-a-star observations is that disk-resolved observations are available that can help interpret the physical causes of spurious RVs (Milbourne et al. 2019, Al Moulla et al. 2023) and provide insights into the potential utility of specific variability indicators (Haywood et al. 2022).

While Sun-as-a-star observations are poised to play a critical role in training, verifying, and validating ML-based models for mitigating solar variability, there is reasonable concern about how well those models can be transferred to mitigate the variability of other stars. Brewer et al. (2020) proposed using stars with giant planets with highly eccentric orbits as RV benchmarks (e.g., HD 3651), since dynamical stability precludes the long-term survival of additional planets over a broad range of orbital periods. Of course, there is uncertainty in the parameters of the giant planet's orbit, and there may still be undetected planets with orbital periods much shorter or longer than the known giant planet. Nevertheless, such stars do appear to be excellent choices for 'standard stars" to be observed intensively by multiple spectrographs, as they are one of the few opportunities to evaluate how well ML-methods trained and tested for the Sun can transfer to other Sun-like stars.†

In practice, astronomers will likely need to rely on a combination of training, testing, validation, and verification on synthetic data, very intensive Sun-as-a-star observations, and less intensive observations of a small number of standard stars. It will be important to formalize how these can be combined to yield robust detections that should be trusted by astronomers and the general public.

In industry, ML methods are commonly evaluated by their predictive power. For many scientific applications, ML algorithms that improve interpretability and explainability can be more valuable than (often more complex) ML algorithms that perform incrementally better. Often, thoughtful choices of features, inspired by physical understanding and/or analysis of synthetic data sets, can allow linear models to perform very nearly as well as more complex and difficult-to-interpret ML methods. While methods such as deep learning are currently quite popular, astronomers should not overlook the power of good feature engineering and more readily interpretable ML methods such as linear models and GPs.

Acknowledgements

I thank numerous collaborators and advisees for their insights, including Debra Fischer, Christian Gilbertson, Zhao Guo, David Jones, Jessi Cisewski-Kehe, Andrea Lin, Thomas Loredo, Jacob Luhn, Suvrath Mahadevan, Michael Palumbo, David Stenning, Alex Wise, Robert Wolpert, Jason Wright, Jinglin Zhao, Lily Zhao, the NEID Science Team, the NASA EPRV RCN, and participants in Astrostatistics workshops organized by the Statistical and Applied Mathematical Sciences Institute. E.B.F. acknowledges partial supported by Heising-Simons Foundation Grant #2019-1177 and NASA EPRV grant #80NSSC21K1035. The Center for Exoplanets and Habitable Worlds supported by Penn State and the Eberly College of Science. The citations in this work have made use of NASA's Astrophysics Data System Bibliographic Services.

† Another possibility is to test methods on near, highly active stars for which interferometric observations are able to constrain the pattern of active regions on the surface (Roettenbacher et al. 2022). However, it is unclear whether the variability seen on active stars will be effective for testing ML-based methods designed for RV quiet stars to be targeted by surveys for Earth-mass planets.

References

Aigrain, S., et al., 2012, MNRAS, 419, 3147. doi:10.1111/j.1365-2966.2011.19960.x
Aigrain, S., Foreman-Mackey, D., 2022, arXiv, arXiv:2209.08940. doi:10.48550/arXiv.2209.08940
Al Moulla, K., et al., 2022, A&A, 664, A34. doi:10.1051/0004-6361/202243276
Al Moulla, K., et al., 2023, A&A, 669, A39. doi:10.1051/0004-6361/202244663
Artigau, É., et al., 2022, AJ, 164, 84. doi:10.3847/1538-3881/ac7ce6
Barragán, O., et al., 2022, MNRAS, 509, 866. doi:10.1093/mnras/stab2889
Bedell, M., et al., 2019, AJ, 158, 164. doi:10.3847/1538-3881/ab40a7
Blackman, R. T., et al., 2020, AJ, 159, 238. doi:10.3847/1538-3881/ab811d
Bouchy, F., et al., 2001, A&A, 374, 733. doi:10.1051/0004-6361:20010730
Brewer, B. J., Stello, D., 2009, MNRAS, 395, 2226. doi:10.1111/j.1365-2966.2009.14679.x
Brewer, J. M., et al., 2020, AJ, 160, 67. doi:10.3847/1538-3881/ab99c9
Cale, B. L., et al., 2021, AJ, 162, 295. doi:10.3847/1538-3881/ac2c80
Camacho, J. D., et al., 2023, MNRAS, 519, 5439. doi:10.1093/mnras/stac3727
Cegla, H. M., et al., 2019, ApJ, 879, 55. doi:10.3847/1538-4357/ab16d3
Chaplin, W. J., et al., 2019, AJ, 157, 163. doi:10.3847/1538-3881/ab0c01
Collier Cameron, A., et al., 2019, MNRAS, 487, 1082. doi:10.1093/mnras/stz1215
Collier Cameron, A., et al., 2021, MNRAS, 505, 1699. doi:10.1093/mnras/stab1323
Crass, J., et al., 2021, arXiv, arXiv:2107.14291. doi:10.48550/arXiv.2107.14291
Cretignier, M., et al., 2020, A&A, 633, A76. doi:10.1051/0004-6361/201936548
Cretignier, M., et al., 2022, A&A, 659, A68. doi:10.1051/0004-6361/202142435
Cretignier, M., et al., 2021, A&A, 653, A43. doi:10.1051/0004-6361/202140986
Davis, A. B., et al., 2017, ApJ, 846, 59. doi:10.3847/1538-4357/aa8303
de Beurs, Z. L., et al., 2022, AJ, 164, 49. doi:10.3847/1538-3881/ac738e
Delisle, J.-B., et al., 2022, A&A, 659, A182. doi:10.1051/0004-6361/202141949
Dodson-Robinson, S. E., et al., 2022, AJ, 163, 169. doi:10.3847/1538-3881/ac52ed
Dravins, D., et al., 2021, A&A, 649, A17. doi:10.1051/0004-6361/202039997
Dumusque, X., et al., 2011, A&A, 525, A140. doi:10.1051/0004-6361/201014097
Dumusque, X., 2018, A&A, 620, A47. doi:10.1051/0004-6361/201833795
Dumusque, X., et al., 2014, ApJ, 796, 132. doi:10.1088/0004-637X/796/2/132
Dumusque, X., et al., 2021, A&A, 648, A103. doi:10.1051/0004-6361/202039350
Dumusque, X., et al., 2015, ApJL, 814, L21. doi:10.1088/2041-8205/814/2/L21
Ervin, T., et al., 2022, ascl.soft, ascl:2207.009. doi:
Fischer, D. A., et al., 2016, PASP, 128, 066001. doi:10.1088/1538-3873/128/964/066001
Ford, E. B., 2006, ApJ, 642, 505. doi:10.1086/500802
Foreman-Mackey, D., et al., 2017, AJ, 154, 220. doi:10.3847/1538-3881/aa9332
Foreman-Mackey, D., et al., 2021, JOSS, 6, 3285. doi:10.21105/joss.03285
Gibson, N. P., et al., 2012, MNRAS, 419, 2683. doi:10.1111/j.1365-2966.2011.19915.x
Gilbertson, C., et al., 2020a, ApJ, 905, 155. doi:10.3847/1538-4357/abc627
Gilbertson, C., et al., 2020b, RNAAS, 4, 59. doi:10.3847/2515-5172/ab8d44
Gilbertson, C., et al., 2024, arXiv, arXiv:2408.17289. doi:10.48550/arXiv.2408.17289
Gully-Santiago, M., Morley, C. V., 2022, ApJ, 941, 200. doi:10.3847/1538-4357/aca0a2
Hall, R. D., et al., 2018, MNRAS, 479, 2968. doi:10.1093/mnras/sty1464
Hara, N. C., Ford, E. B., 2023, AnRSA, 10, 623. doi:10.1146/annurev-statistics-033021-012225
Haywood, R. D., et al., 2022, ApJ, 935, 6. doi:10.3847/1538-4357/ac7c12
Haywood, R. D., et al., 2016, MNRAS, 457, 3637. doi:10.1093/mnras/stw187
Haywood, R. D., et al., 2014, MNRAS, 443, 2517. doi:10.1093/mnras/stu1320
Holzer, P. H., et al., 2021, AJ, 161, 272. doi:10.3847/1538-3881/abf5e0
Jones, 2022, Ann. Appl. Stat. 16, 652-679. doi:10.1214/21-AOAS1471
Jurgenson, C., et al., 2016, SPIE, 9908, 99086T. doi:10.1117/12.2233002
Langellier, N., et al., 2021, AJ, 161, 287. doi:10.3847/1538-3881/abf1e0
Lin, A. S. J., et al., 2022, AJ, 163, 184. doi:10.3847/1538-3881/ac5622
Luhn, J. K., et al., 2020, AJ, 159, 235. doi:10.3847/1538-3881/ab855a
Luhn, J. K., et al., 2023, AJ, 165, 98. doi:10.3847/1538-3881/acad08

Meunier, N., et al., 2015, A&A, 583, A118. doi:10.1051/0004-6361/201525721
Meunier, N., et al., 2010, A&A, 512, A39. doi:10.1051/0004-6361/200913551
Milbourne, T. W., et al., 2019, ApJ, 874, 107. doi:10.3847/1538-4357/ab064a
Newman, P. D., et al., 2023, AJ, 165, 151. doi:10.3847/1538-3881/acad07
Nicholson, B. A., Aigrain, S., 2022, MNRAS, 515, 5251. doi:10.1093/mnras/stac2097
Palumbo, M. L., et al., 2022, AJ, 163, 11. doi:10.3847/1538-3881/ac32c2
Perger, M., et al., 2023, arXiv, arXiv:2301.12872. doi:10.48550/arXiv.2301.12872
Petersburg, R. R., et al., 2020, AJ, 159, 187. doi:10.3847/1538-3881/ab7e31
Queloz, D., et al., 2001, A&A, 379, 279. doi:10.1051/0004-6361:20011308
Rajpaul, V., et al., 2016, MNRAS, 456, L6. doi:10.1093/mnrasl/slv164
Rajpaul, V., et al., 2015, MNRAS, 452, 2269. doi:10.1093/mnras/stv1428
Rajpaul, V. M., et al., 2020, MNRAS, 492, 3960. doi:10.1093/mnras/stz3599
Roettenbacher, R. M., et al., 2022, AJ, 163, 19. doi:10.3847/1538-3881/ac3235
Silva, A. M., et al., 2022, A&A, 663, A143. doi:10.1051/0004-6361/202142262
Wise, A. W., et al., 2018, AJ, 156, 180. doi:10.3847/1538-3881/aadd94
Wise, A., et al., 2022, ApJ, 930, 121. doi:10.3847/1538-4357/ac649b
Zechmeister, M., et al., 2019, A&A, 627, A49. doi:10.1051/0004-6361/201935460
Zechmeister, M., et al., 2018, A&A, 609, A12. doi:10.1051/0004-6361/201731483
Zhao, J., et al., 2022, ApJ, 935, 75. doi:10.3847/1538-4357/ac77ec
Zhao, L., et al., 2020, RNAAS, 4, 156. doi:10.3847/2515-5172/abb8d0
Zhao, L. L., et al., 2022, AJ, 163, 171. doi:10.3847/1538-3881/ac5176
Zhao, Y., Dumusque, X., 2023, A&A, 671, A11. doi:10.1051/0004-6361/202244568

Machine Learning in Astronomy: Possibilities and Pitfalls
Proceedings IAU Symposium No. 368, 2025
J. McIver, A. Mahabal & C. Fluke, eds.
doi:10.1017/S1743921323001394

How do we design data sets for Machine Learning astronomy?

Renée Hložek[1,2]

[1]Dunlap Institute for Astronomy & Astrophysics, 50 St. George Street, Toronto ON M5S 3H4

[2]David A. Dunlap Department for Astronomy & Astrophysics, 50 St. George Street, Toronto ON M5S 3H4

email: hlozek@dunlap.utoronto.ca

Abstract. Many problems in astronomy and physics lend themselves to solutions from machine learning methods for the detection and classification of astronomical signals, and model inference from those signals. The historic presentation of machine learning methods as 'black boxes' has generated push back from some in the the physics/astronomy communities regarding how useful they are to truly uncover the physical laws that govern our world. Skepticism about the applicability of new computational methods in scientific inference is not new; we highlight connections between the machine learning contexts and previous computational paradigm shifts in astronomy. Moreover, several advances in methodologies challenge the assumption that machine learning 'gives us answers that we can use but do not understand' to standing physics questions. We summarize some astronomical machine learning data challenges used in astronomy and how we can use challenges on different scales to test different parts/use cases of our analysis methods.

Keywords. machine learning, surveys, observations, statistics

1. Introduction

We are firmly in the artificial intelligence/machine-learning (AI/ML) assisted age. As the volume of data in astronomy increases to the 'petabyte age', astronomers are constantly looking for methods to process raw data to compressed science-ready products and discard data that are not useful for astronomical inference. Advances in the complexity and robustness of machine learning methods have been equally matched with an increase in their use in astronomy and physics contexts.

1.1. *Computational advances lead to both innovation and skepticism*

Machine learning is currently a widely used tool in astronomical research, with advances in algorithm development and methodology driving rapid breakthroughs in previously intractable problems. As breathtaking as this current paradigm is, it mirrors a similar 'revolution' in astronomy: that of simulating large gravitational systems in the late 1960s and 1970s. Advances in computing led to an explosion of the so-called 'N-body' simulations of gravitation. As shown in Figure 1, following their initial use, publications using the phrase 'N-body' increased dramatically following the similar increase in computational power. It is sometimes taken for granted that N-body simulations are an excellent way to study the dynamics of gravitational systems, however when they were first developed there was real skepticism of the validity of the methods and the divergence

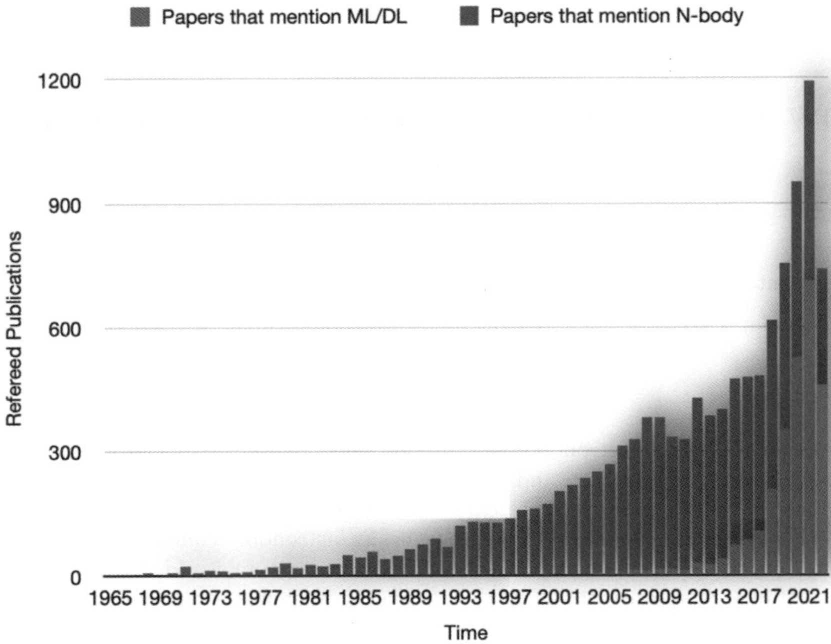

Figure 1. The number of refereed publications mentioning artificial intelligence/machine learning (pink) compared to those mentioning N-body methods (purple) as a function of time returned by the NASA Astrophysics Data System online at NASA ADS. The AI/ML revolution can be seen as the rapid growth in the number of publications as a function of time.

between groups following different prescriptions. Similarly, simulations diverged rapidly even with tiny changes in the initial conditions, as noted in early studies Miller (1964, 1971).

In a meeting to discuss these new methods, M. Lecar issued a word of caution and suggested the need for further comparative studies, given the one attempted for the 1968 meeting of researchers in dynamics:

It seems that numerical experiments are becoming increasingly popular, and before things get out of hand, I feel that further tests of reproducibility are imperative. However, because of the difficulties in handling volumes of data from different computers, it would be worthwhile for each of us first to initiate statistical studies of our own; varying the accuracy and the microscopic initial conditions, and searching for stable quantities. These results should be sent to all the participants in this study, and perhaps by next summer, we can hold a second comparison study. Lecar (1968)

As summarized in Quinlan & Tremaine (1992), this sensitivity of N-body simulations to small errors was not simply a numerical artifact that would dissolve with additional computational power, and so the community had to establish (and compare) best practices for how to regularize the integration methods in these systems. The community did not, however, abandon the N-body methods due to these potential pitfalls. Comparisons across groups on 'standard systems', clear descriptions of methods and exploration of the impacts of methodological assumptions has led to the continued success of this technique in astronomy for systems across a range of scales.

1.2. *What is machine learning in astronomy used for?*

Machine learning methods in astronomy have been used since the mid-late 1990s (see Baron 2019, for an excellent comparison of different algorithms and their application in astronomy), with early methods focusing on spectral analysis (Connolly et al. 1995; Boroson & Green 1992) and principal component analysis slowly replaced with more recent use cases including deep learning, neural networks and diffusion methods (Cireşan et al. 2012; Fukushima 1980; Ho et al. 2020; Ćiprijanović et al. 2022). The use cases in astronomy can be summarized in roughly three categories, namely the *detection* of signals in a faint and often varying background; the *clustering/classification* of objects into various categories through either unsupervised (clustering) or supervised (classification) methods; and statistical *inference* of certain parameters or models allowed by a given data set.

1.2.1. Signal detection

Astronomical signals can vary from the 'extremely bright/loud' bursts of radiation to faint hints of signal within a bright background. Examples of bright objects include Fast Radio Bursts (FRBs), which exhibit a characteristic 'sweep' in frequency-time space as the signal passes through the electrons along the line of sight. First detected in 2007 (Lorimer et al. 2007), fast radio bursts are exciting new objects in the sky that display brighten and fade rapidly in radio bands (ranging from milliseconds to a few seconds in duration). Some objects repeat this brightening/fading, while others appear to brighten only once (The CHIME/FRB Collaboration et al. 2023). Several surveys are optimized to find many of these new objects in the hopes of uncovering their origin and the physical mechanism behind the bursts (CHIME/FRB Collaboration et al. 2021; Connor et al. 2023; Vanderlinde et al. 2019; Lin et al. 2022; van Leeuwen et al. 2023; Megias Homar et al. 2023). The gravitational wave signatures of the in-fall of pairs of massive bodies like black holes and neutron stars also have a characteristic shape, but are a much fainter signal relative to background effects. First detected in 2015 (Abbott et al. 2016), these signals are powerful probes of the model of General Relativity and paired with electromagnetic counterparts, they offer a window into the energetics of these powerful events. Detecting the signals above the background and/or distinguishing the signal from an instrumental artifact is an active area of research (Alvarez-Lopez et al. 2023; Bini et al. 2023; George & Huerta 2018b,a; Razzano et al. 2023; Wang et al. 2023; Cabero et al. 2020). Assigning an astrophysical host to both fast radio bursts and gravitational wave sources is complicated by the uncertainty in spatial localization of the signal and the density of galaxies in the sky.

1.2.2. Clustering and classification

Once astronomical signals have been detected, a separate goal is to separate the signals into different classes (or sub-classes) of objects. Supervised approaches use sets of real or simulated data with labels to train machine learning algorithms to recognize characteristic features in the data. These data typically contain changing flux in a given electromagnetic band (optical, radio, X-ray etc.) as a function of time. In the case of FRBs, various machine learning methods have been developed to classify these objects and distinguishing them apart from radio frequency interference (RFI) signals present in the data, including supervised (Connor & van Leeuwen 2018; Zhang et al. 2020; Agarwal et al. 2020) and unsupervised methods Zhu-Ge et al. (2023). In the case of the gravitational wave data, recent approaches separate the detected signals into different classes based on models of distances to the detected signals (T. C. Abbott et al. 2022).

In both cases, very rapid follow-up of sources by complementary telescopes maximizes the scientific return of the observation, and so classifiers need to return results in a short period of time. Similar time constraints exist for the classification of events in gamma ray and X-ray telescopes, where detections of interest need to be flagged and communications sent out to other telescopes for 'target-of-opportunity' observations (Tohuvavohu et al. 2020; DeLaunay & Tohuvavohu 2022).

In optical astronomy, time-series classification and clustering has enjoyed a longer history of development. In this case detection itself is less of a challenge as the signals typically trigger a discovery 'alert' depending on a brightness threshold relative to the background noise. 'Forced' or archival photometric data are then produced from observations of the same location on the sky before the outburst/detection to determine the pre-outburst brightness of the host or environment.

1.2.3. Inference

While ML algorithms can provide predictive results for the detection or classification of new signals within noisy data, physicists are driven by the need to understand underlying processes that govern the behaviour of stars, galaxies, planets and the cosmos. In order to infer parameters of the model and uncertainties, very different machine learning architectures must be used. One method used to perform inference using ML is using Bayesian Neural Networks. In this formalism, weights and biases attached to the neural network are not specified as individual parameters but with probability distributions. The result of these distributions is that they can yield *post facto* distributions in the learned parameters of the network, providing model output that is more interpretable and transferable. While the approach of adding uncertainty to model networks was around from the early 1990s (Denker & LeCun 1990; Buntine & Weigend 1991; MacKay 1992), the application to modern machine learning gathered steam in recent years (Blundell et al. 2015) and is now commonly used in astrophysics to build models of galaxy formation and morphology (Walmsley et al. 2019; Dunn et al. 2023; Reza et al. 2022; Piras et al. 2023; Lucie-Smith et al. 2023), supernova modelling (Stein et al. 2022) and large-scale structure clustering (Sullivan et al. 2023; Modi et al. 2021), to name but a few examples.

2. Some Pitfalls/Critiques of Machine learning in Astronomy

In order to take full advantage of the machine assisted revolution in astronomy, we need to understand and address some of the critiques leveled at the use of machine learning in astronomy. We list some of the common pitfalls and describe attempts by the community to address them.

2.1. *Machine Learning provides 'black boxes' without any physics understanding*

The use of machine learning in astronomy/physics is often criticized for being a 'black box' that is not understood, or that parts of the model are not transparently discussed. As Rudin (2018) defines it, this black box could be because the model is a function that is too complicated (e.g. recursive) for any human to comprehend or that the function contains parts that are proprietary (something more common in medical, rather than astronomical applications). A key distinction that Rudin makes is between the *explainability* of the black box/algorithm, and the *interpretability* of the model. It is the latter that is of most interest in physics, as whatever model is used, we want it to illuminate wherever possible the underlying physical processes that the system is trying to model. Moreover, some astronomers see machine learning models as tools to use within the context of their analysis, and use off-the-shelf methods and tools to ensure speed and robustness of their

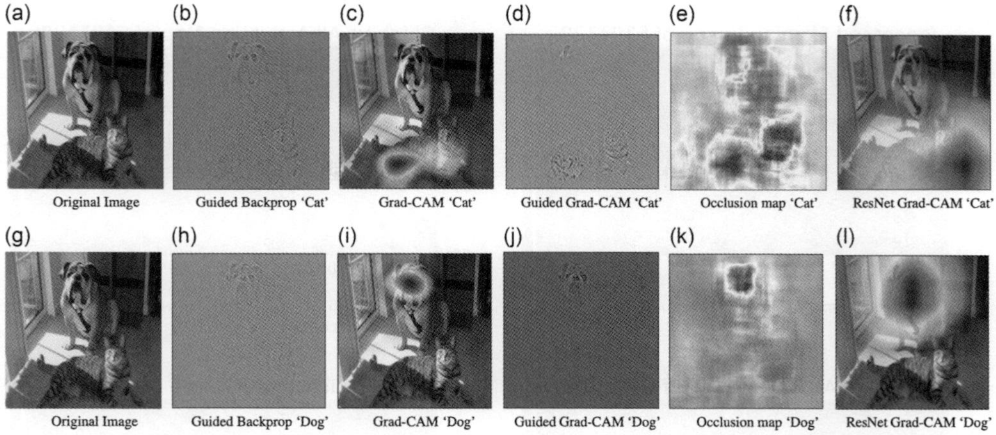

Figure 2. The Grad-CAM algorithm uses a class activation map weighted by the gradient information flowing into the final layer of a CNN to determine which parts of the data (in this case an image) are yielding the classification of 'cat' or 'dog'. Figure reproduced from (Selvaraju et al. 2016).

analysis pipelines. As such, less time is often spent on either justifying the approach used (or the selection of the network architecture) and explaining why it is useful in the specific astronomical context. This 'means-to-an-end' approach can perpetuate the myth that these tools lack value or that they do not allow for deeper understanding of the physics.

As outlined in Grojean et al. (2022), some methods have been introduced to develop *post facto* interpretability of trained ML models, including 'local' methods that trace every outcome in terms of all the input variables provided (e.g. LIME, Tulio Ribeiro et al. 2016), or more global methods analyse which variables are the most 'important' to the algorithm, or that produce the most variation of output through variables like the Gini coefficient (Giorgi & Gigliarano 2017).

In order to understand what their machine-learning algorithms are learning (or rather, what in the data they are responding to), methods like Gradient-weighted Class Activation Mapping (Grad-CAM, Selvaraju et al. 2016) to determine what parts of an image determine the classification/typing of the image. Grad-CAM is based on the Class Activation Mapping (CAM, Zhou et al. 2015) method which creates class activation maps by using the global average pooling performed in Convolutional Neural Networks (CNNs Fukushima 1980) to indicate parts of the image associated with the target category. Illustrated in Figure 2, the Grad-CAM algorithm modifies this by using the gradient information flowing into the final convolutional layer of the network as a measure of the importance of each neuron to the particular region of the image/data in determining the classification. This method is more robust to the underlying CNN and can provide useful insight into what features your machine learning method is actually learning. This has successfully been applied to the astronomical context in determining what the ML algorithms were learning when determining whether a gravitational wave signal was astrophysical or an instrumental glitch (T. C. Abbott et al. 2022, as shown in Figure 3).

Others use the approach of using regressive algorithms to 'directly infer' physical laws themselves. Schmidt & Lipson (2009) suggested that they could recover natural physical laws from observations alone by using machine learning methods, however further inspection by Hillar & Sommer (2012) suggested that their fitness functions within the algorithm incorporated Hamilton's equations of motions and Newton's second law

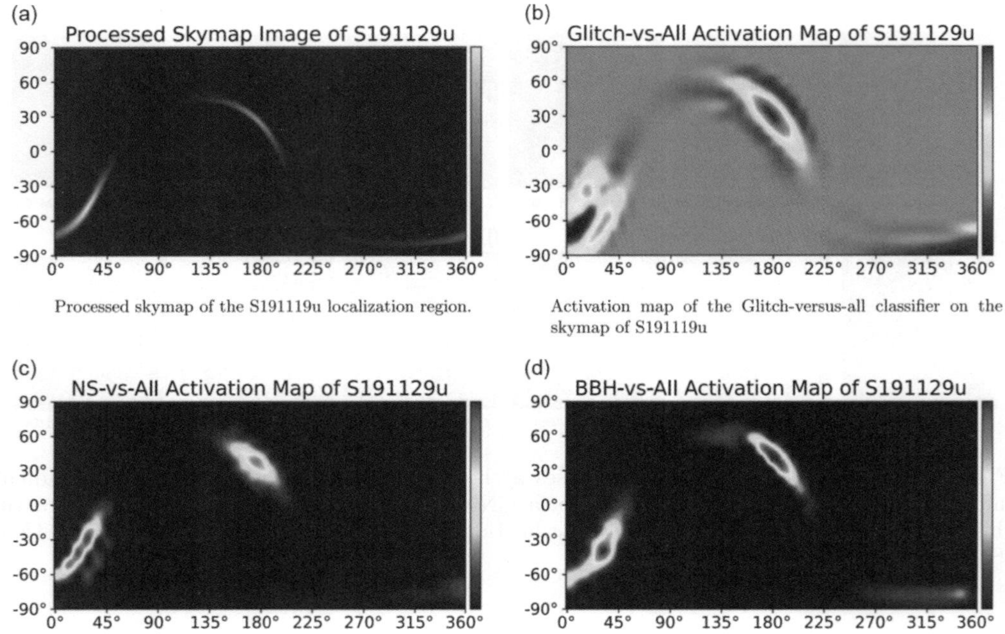

(a) Processed skymap of the S191119u localization region.

(b) Activation map of the Glitch-versus-all classifier on the skymap of S191119u

(c) Activation map of the NS-versus-all classifier on the skymap of S191129u.

(d) Activation map of the BBH-versus-all classifier on the skymap of S191129u

Figure 3. Grad-CAM as applied to `GWSkyNet` classifications of binary black hole or neutron star pairs. Figure reproduced from (T. C. Abbott et al. 2022).

directly. Hillar & Sommer (2012) suggest ways to enforce a general fitness function that is not prone to this bias.

In the past decade, physics-informed machine learning models (Raissi et al. 2019) incorporate physical laws (for example those that govern the time-dependent dynamics of the system or those that control the energy budget) directly into the algorithm as prior information, rather than applying a generic machine-learning algorithm to the problem and assuming no prior information. Including these laws as rules into the learning algorithm itself can lead to more efficient and robust algorithm performance on complex problems in physics, and can reduce the dependence on training, or allow unsupervised training altogether. See (Karniadakis et al. 2021) for a review of these methods and their applicability to problems in physics and mathematics.

Applying these physics-aware or physics-based machine learning approaches to solve problems in physics and astronomy is a growing area of study and has particular applicability in the study of nonlinear dynamics and fluid mechanics (see e.g. Shukla et al. 2022; Dai et al. 2023; Rosofsky et al. 2022; Karpov et al. 2022; Iyer et al. 2022; Lucie-Smith et al. 2023; Ntampaka et al. 2021). A challenge suggested in Nord et al. (2019) is to facilitate the use of these physics-informed algorithms to solve some of the large and complex modeling required in the current age of large astronomical data.

2.2. *Machine learning does not provide error estimates*

Another criticism leveraged against machine learning methods historically has been a lack of meaningful error estimates from 'black box' models, however the past decade has seen the introduction of ML approaches specifically designed to provide uncertainty estimates on the parameters of interest within a model. These innovations were developed to enable machine learning *inference* of the model parameters (a few examples of current uses of inference in astronomy are included above). In fact, there is a move in the community to compare different approaches of uncertainty quantification (UQ) in physics and astronomy (see Caldeira & Nord 2020; Psaros et al. 2023; Ćiprijanović et al. 2022; Mohan et al. 2022; Ntampaka & Vikhlinin 2022, for some recent examples).

Machine-learning approaches and methods are being developed at a somewhat faster pace than the general astronomical community up-take. However, community standards continue to evolve towards these more interpretable approaches. Astronomy research planning efforts in e.g. the United States and Canada occur every decade and included these planning efforts are white papers and discussions on machine learning trends, resources and challenges – provided a natural platform for the discussions to move the field towards methods that can provide uncertainty estimates, are interpretable and physically motivated (e.g. Nord et al. 2019; Dvorkin et al. 2022; Siemiginowska et al. 2019).

2.3. *Machine learning compresses (and loses) information*

In order to process large data sets through machine learning algorithms, these data are often 'compressed' onto a smaller set of features/axes/vectors of information. While this can be criticized as 'losing information' in the system (e.g. by splitting up a video into many still frames, thereby potentially losing the temporal connection between successive frames), it is also important to address the problem of over-fitting in ML analyses. This occurs when the ML model starts fitting to noise in the data set, given enough model flexibility.

Deep learning methods were developed specifically to handle very large and complex data sets (e.g. 2D images and human speech) where there is a risk of under-fitting/not building a complete and predictive model. Changes to model complexity and network architecture can be made to ensure that the model does not have too much flexibility and freedom if there is concern of overfitting noise. Here we can learn from colleagues in optics and machine vision, who investigate when (and what) data to add into a system to obtain model that is optimally fit and that remains predictive (Boulahia et al. 2021). The challenge remains for the astronomy community to test and review any ML framework used for a specific analysis rather than dismissing the tools themselves. Some groups are already thinking at this abstract level and evaluating robustness, (Ćiprijanović et al. 2022), and examining how we can generalize the generating and simulating of data sets for such machine-learning contests (Lewis et al. 2022).

3. Leveraging data challenges in Astronomy machine learning

Astronomy has a rich history of simulating data sets in advance of telescope operations, and through regular advances in telescope development, has been in a 'data-rich' paradigm for many years. While the era of 'data-deluge' from large survey instruments like the Vera C. Rubin Observatory (Rubin Obs, LSST Science Collaboration et al. 2009; Bianco et al. 2022; LSST Dark Energy Science Collaboration 2012) and the Square Kilometer Array (SKA, Square Kilometre Array Cosmology Science Working Group et al. 2020) will soon be upon us, using current surveys (see e.g. van Roestel et al. 2021; T. M. C. Abbott et al. 2022) and simulations of the sky we will observe with these new facilities (LSST Dark Energy Science Collaboration (LSST DESC) et al. 2021; Levrier et al. 2009; Ramírez-Pérez et al. 2022). This ability to simulate, and to test pipelines on existing data sets with 'known' ground truth, and the community standard of transferring methods to different observing conditions and telescope configurations is a competitive advantage in astronomy, and one that distinguishes us from other data-rich fields where simulations are either extremely costly to obtain, complicated, or where data are highly individualized due to experimental conditions (An 2018; Klingner et al. 2022; Errington et al. 2021).

What makes an effective astronomical machine-learning data challenge? Boscoe et al. (2022) describe how effective astronomical data challenges need to have well-defined data points (including only what is relevant to the challenge, ensuring there are tags for data quality, including well-modeled uncertainties) and even make recommendations for the structural form of data (HDF5 rather than the common FITS format that is so common in astronomy. In addition to the data format, it is important to have well-structured and easy to access metadata, to enable supervised learning without the full domain knowledge that is sometimes mistakenly assumed of astronomers participating in these challenges. These recommendations for the well-defined data, data-structure and metadata can be encapsulated in the recommendation that the data challenge *goals are clearly defined* and articulated to participants, that *any required domain knowledge is communicated* and that the *performance metrics are well-established and communicated.* This can be challenging given the multi-variate needs of the astronomical community interested in a particular data challenge, however the preparatory work in defining goals, assumed knowledge and metrics can greatly change the number of potential participants and the success of the challenge. We highlight a few astronomical challenges below.

3.1. *Time-series classification challenges*

The sky has some objects in it (galaxies, groups and clusters of galaxies) that change slowly with time, other objects that fluctuate in brightness (e.g. energetic jets from the centre of galaxies, spinning neutron stars or pulsating stars nearing stellar death) and 'new' or transient objects that are a result of explosions, the collision of pairs of massive objects or other undefined phenomena. Classifying these transient objects into different categories is needed in order to analyze specific samples/types, and astronomers have been developing tools for rapid classification in tandem with the increase in the survey speed of new telescopes. Early surveys like the Catalina Real-Time Transient Sky Surveys (Drake et al. 2012) and the Palomar-Quest Digital Synoptic Sky Survey (Djorgovski et al. 2008) ushered in an era of wide-field transient science and presented lessons learned for processing of transient notifications or 'alerts' (see e.g. Graham et al. 2012). The Sloan Digital Sky Survey (Kessler et al. 2009; Sako et al. 2018) was focused on developing a sample of Type Ia supernovae (SNIa) for cosmological analysis and developed methods to classify the observed transients into different types to facilitate spectroscopic follow up of likely SNIa candidates.

Recognizing the challenges that this classification had presented, Kessler *et al.* developed the Supernova Photometric Classification Challenge (SNPhotCC, Kessler et al. 2010) which was a classification challenge based on simulated supernovae of three different types assuming the photometric properties of a survey like the (then upcoming) Dark Energy Survey. The challenge led to broad community participation (Kessler et al. 2010) and engaged new groups in optical classification. The challenge highlighted that the community particularly struggled with the increase of data (and increase in noise) with survey depth, and that the change in relative numbers of different populations when going from precursor data to larger volume data sets. As a result, when simulating a new survey for the upcoming Rubin Observatory, the Photometric LSST Astronomical Time-series Classification Challenge (PLAsTiCC, Kessler et al. 2019; Hložek et al. 2020; Malz et al. 2018) included many more types of transients, a large difference in the size of the training sample relative to the test sample (as shown in Figure 4), and a collection of rare objects that were explicitly excluded from any labelled training data. The three-month PLAsTiCC (hosted on a public data challenge website at Kaggle.com) had over 1000 participants from within and outside the astronomical community and has

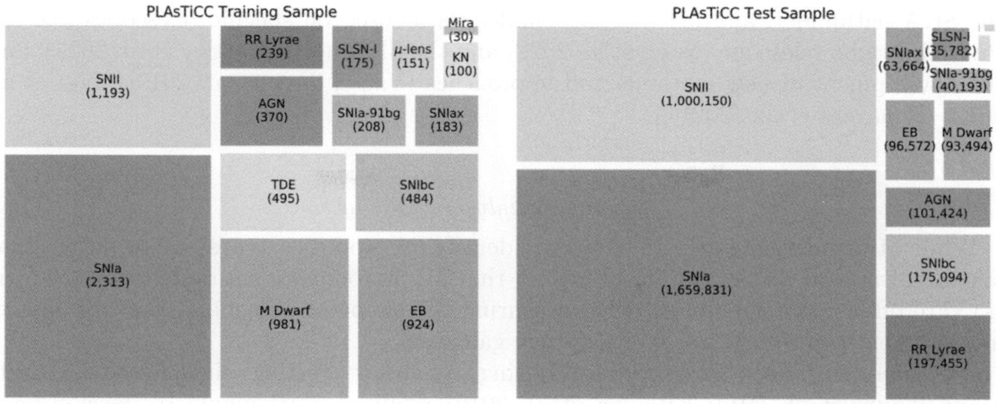

Figure 4. The $\simeq 8000$ objects in the training set and 3.5 million test objects in the PLAsTiCC survey. The size of the boxes are proportional to the relative numbers in each set and the absolute numbers are in parentheses. Figure reproduced from Hložek et al. (2020).

already led to 45 publications on methods and approaches to classification and anomaly detection.

This challenge included time-series data for the full LSST survey from Rubin, as processed flux measurements. The primary way that new transients will be communicated from the telescope, however, is through packages of 'alerts' released by the telescope to a specific set of alert brokers developed within the community. The next instalment in developing classification challenges was to provide the light curves in this 'alert' format, and to increase the complexity of the simulations to control for host environment (Lokken et al. 2023). The Extended LSST Astronomical Time-series Classification Challenge (ELAsTiCC, Narayan & ELAsTiCC Team 2023) was focused at these brokers, testing their ability to ingest and process (and classify) a stream of alerts similar to what will be generated from Rubin. The challenge is available online.

PLAsTiCC and ELAsTiCC are broad challenges including a range of different object types. Narrower, domain-specific classification challenges can include more model variety and employ more specific science-focused metrics for evaluating performance between methods. The LSST AGN classification challenge (Yu et al. 2022) provided mock observations based on templates from existing data and focused on assessing the ability of participants to select AGN and parameterize their light-curves, and the ability to determine the photometric redshift of the AGN candidate.

While not considered 'transient', finding planets around other stars requires similar techniques as the exoplanets often transit their host stars, or generate faint shifts to the motion of the host star. The Radial Velocity Challenge (Dumusque et al. 2017) simulated radial velocity signals (Dumusque 2016) from the 'wobble' that an exoplanet would generate for its host star. The challenge was focused on testing methods in advance of the Transiting Exoplanet Survey Satellite (TESS, Ricker et al. 2014).

The Laser Interferometer Space Antenna (LISA) Data challenges (Baghi 2022) are a series of challenges designed to test community methods for extracting gravitational wave signals from noisy data in preparation for the European Space Agency (ESA)- and NASA-funded LISA mission, with the current purpose of tackling mild source confusion within an idealized instrumental noise framework. The current challenge continues a long line of previous challenges developed by the Mock LISA Data Challenge task force (Babak et al. 2008) to prepare the community for LISA analysis.

Similarly, the Australian Square Kilometre Array Pathfinder (ASKAP) group designed a data challenge to test the fidelity of radio source finding (Hopkins et al. 2015) for the SKA pathfinder, and led to other challenges focused on specific telescopes like the PARKES radio telescope (Yong et al. 2022) and the SKA itself (Hartley et al. 2023), and the development of new methods and approaches (Vafaei Sadr et al. 2019; Riggi et al. 2019; Bonavera et al. (2021).

3.2. *Image analysis challenges*

While transient surveys focus on identifying specific types of objects that appear/disappear or vary in brightness on the sky, astronomers interested in the 'static sky' are often more concerned with measuring the shape of objects, classifying objects into different types (e.g. spiral or elliptical galaxies).

For example, the GRavitational lEnsing Accuracy Testing challenges (GREAT, Mandelbaum et al. 2014; Kitching et al. 2010) required participants to develop lensing shear estimators and compared performance on increasingly complex simulated test data. The LSST-DESC 3x2pt Tomography Optimization Challenge (Zuntz et al. 2021), was challenge with a more specialized purpose: to determine the optimal tomographic binning scheme for a photometric survey for measurements of the '3x2pt' (the three different types of two-point correlation function measurements used as the data observable that summarizes the clustering of galaxies, and the shapes of those galaxies as they are distorted by cosmic lensing.) This challenge, targeted at members of the LSST Dark Energy Science Collaboration (DESC) was a way to galvanize the DESC community to generating an optimal method of defining samples for the upcoming Rubin LSST survey.

The above lensing challenges focus on the subtle affects from weak gravitational lensing. 'The Strong Gravitational Lens Finding Challenge' (Metcalf et al. 2018) instead asked participants to identify 100,000 candidate strong lens systems from millions of noisy images. By comparing algorithm performance across all entries, the challenge presented also gained new insights e.g. for the need for multi-band imaging in order to correctly identify strong lens systems.

3.3. *The power of citizen science*

While many challenges in astronomy are reasonably technical and can require some amount of domain knowledge, large-scale challenges that take advantage of the energy of 'citizens' can generate large amounts of data and find novel objects. One of the classic examples of citizen science is the Galaxy Zoo project (Lintott et al. 2008), which started off asking citizens to make simple classifications between spiral and elliptical galaxies in SDSS images, but has grown to encompass citizen science challenges across disciplines (see the full suite at zooniverse.org) and led to the discovery in astronomy of novel objects like the 'green pea' galaxies, named after their greenish appearance and small size in the SDSS images. Further study of these new objects discovered by the citizen scientists revealed their role in the reionization of the universe (Cardamone et al. 2009). The classifications from citizen science volunteers can then be used to train and refine machine-learning methods for discovery and classification (Andersson et al. 2023; Peek & White 2021; Kruk & Merín 2023; Razzano et al. 2023; Jimenez et al. 2023).

4. Conclusions

Data challenges are an important step in preparing for the complexity of future data sets, and for developing new approaches to analyze current data. Astronomers have the benefit of extensive and well-labelled existing data sets, and the community practice of simulating new data with different intrinsic properties and observing conditions. When developing a data challenge for training/testing machine learning methods in astronomy, the first step should be reducing and controlling the complexity of the data presented, and deciding on the limited number of goals of the challenge.

These astronomical challenges provide great ways of galvanizing the astronomical community to develop new methods, and if well-curated the data set will remain useful well beyond the individual data challenge. It may be that data challenges need to be developed in stages, increasing the complexity of the data used in each stage, and relaxing additional assumptions about metadata or data complexity with each challenge. Rather than seeing this as a bug, it can be approached as a smooth way to understand the community response to data complexity.

A challenge that remains for the community is to include rewards for methods that are physically interpretable, and to develop standards for how to communicate and derive machine learning solutions to problems in astronomy. We are lucky as a community to continually have exciting new data on the horizon, and look forward to more machine-assisted solutions to understand the Universe.

References

Abbott, B. P., et al. (2016). Tests of general relativity with GW150914. *Phys. Rev. Lett.*, *116*(22), 221101. ([Erratum: Phys.Rev.Lett. 121, 129902 (2018)]) doi: 10.1103/PhysRevLett.116.221101

Abbott, T. C., Buffaz, E., Vieira, N., Cabero, M., Haggard, D., Mahabal, A., & McIver, J. (2022, March). GWSkyNet-Multi: A Machine-learning Multiclass Classifier for LIGO-Virgo Public Alerts. *ApJ*, *927*(2), 232. doi: 10.3847/1538-4357/ac5019

Abbott, T. M. C., Aguena, M., Alarcon, A., Allam, S., Alves, O., Amon, A., ... DES Collaboration (2022, January). Dark Energy Survey Year 3 results: Cosmological constraints from galaxy clustering and weak lensing. *Phys. Rev. D*, *105*(2), 023520. doi: 10.1103/PhysRevD.105.023520

Agarwal, D., Aggarwal, K., Burke-Spolaor, S., Lorimer, D. R., & Garver-Daniels, N. (2020, September). FETCH: A deep-learning based classifier for fast transient classification. *MNRAS*, *497*(2), 1661–1674. doi: 10.1093/mnras/staa1856

Alvarez-Lopez, S., Liyanage, A., Ding, J., Ng, R., & McIver, J. (2023, April). GSpyNet-Tree: A signal-vs-glitch classifier for gravitational-wave event candidates. *arXiv e-prints*, arXiv:2304.09977. doi: 10.48550/arXiv.2304.09977

An, G. (2018, September). The crisis of reproducibility, the denominator problem and the scientific role of multi-scale modeling. *Bull Math Biol*, *80*(12), 3071–3080.

Andersson, A., Lintott, C., Fender, R., Bright, J., Carotenuto, F., Driessen, L., ... Whittle, I. (2023, April). Bursts from Space: MeerKAT - The first citizen science project dedicated to commensal radio transients. *arXiv e-prints*, arXiv:2304.14157. doi: 10.48550/arXiv.2304.14157

Babak, S., Baker, J. G., Benacquista, M. J., Cornish, N. J., Crowder, J., Cutler, C., ... Challenge-2 participants, t. (2008, June). Report on the second Mock LISA data challenge. *Classical and Quantum Gravity*, *25*(11), 114037. doi: 10.1088/0264-9381/25/11/114037

Baghi, Q. (2022, April). The LISA Data Challenges. *arXiv e-prints*, arXiv:2204.12142. doi: 10.48550/arXiv.2204.12142

Baron, D. (2019, April). Machine Learning in Astronomy: a practical overview. *arXiv e-prints*, arXiv:1904.07248. doi: 10.48550/arXiv.1904.07248

Bianco, F. B., Ivezić, Ž., Jones, R. L., Graham, M. L., Marshall, P., Saha, A., ... Willman, B. (2022, January). Optimization of the Observing Cadence for the Rubin Observatory Legacy Survey of Space and Time: A Pioneering Process of Community-focused Experimental Design. *ApJS*, *258*(1), 1. doi: 10.3847/1538-4365/ac3e72

Bini, S., Vedovato, G., Drago, M., Salemi, F., & Prodi, G. A. (2023, March). An autoencoder neural network integrated into gravitational-wave burst searches to improve the rejection of noise transients. *arXiv e-prints*, arXiv:2303.05986. doi: 10.48550/arXiv.2303.05986

Blundell, C., Cornebise, J., Kavukcuoglu, K., & Wierstra, D. (2015, May). Weight Uncertainty in Neural Networks. *arXiv e-prints*, arXiv:1505.05424. doi: 10.48550/arXiv.1505.05424

Bonavera, L., Suarez Gomez, S. L., González-Nuevo, J., Cueli, M. M., Santos, J. D., Sanchez, M. L., ... de Cos, F. J. (2021, April). Point source detection with fully convolutional networks. Performance in realistic microwave sky simulations. *A&A, 648*, A50. doi: 10.1051/0004-6361/201937171

Boroson, T. A., & Green, R. F. (1992, May). The Emission-Line Properties of Low-Redshift Quasi-stellar Objects. *ApJS, 80*, 109. doi: 10.1086/191661

Boscoe, B., Do, T., Jones, E., Li, Y., Alfaro, K., & Ma, C. (2022, November). Elements of effective machine learning datasets in astronomy. *arXiv e-prints*, arXiv:2211.14401. doi: 10.48550/arXiv.2211.14401

Boulahia, S. Y., Amamra, A., Madi, M. R., & Daikh, S. (2021, nov). Early, intermediate and late fusion strategies for robust deep learning-based multimodal action recognition. *Mach. Vision Appl., 32*(6). doi: 10.1007/s00138-021-01249-8

Buntine, W. L., & Weigend, A. S. (1991). Bayesian back-propagation. *Complex Syst., 5*.

Cabero, M., Mahabal, A., & McIver, J. (2020, November). GWSkyNet: A Real-time Classifier for Public Gravitational-wave Candidates. *ApJ, 904*(1), L9. doi: 10.3847/2041-8213/abc5b5

Caldeira, J., & Nord, B. (2020, April). Deeply Uncertain: Comparing Methods of Uncertainty Quantification in Deep Learning Algorithms. *arXiv e-prints*, arXiv:2004.10710. doi: 10.48550/arXiv.2004.10710

Cardamone, C., Schawinski, K., Sarzi, M., Bamford, S. P., Bennert, N., Urry, C. M., ... Vandenberg, J. (2009, November). Galaxy Zoo Green Peas: discovery of a class of compact extremely star- forming galaxies. *MNRAS, 399*, 1191–1205. doi: 10.1111/j.1365-2966.2009.15383.x

CHIME/FRB Collaboration, Amiri, M., Andersen, B. C., Bandura, K., Berger, S., Bhardwaj, M., ... Zwaniga, A. V. (2021, December). The First CHIME/FRB Fast Radio Burst Catalog. *ApJS, 257*(2), 59. doi: 10.3847/1538-4365/ac33ab

Ćiprijanović, A., Kafkes, D., Snyder, G., Sánchez, F. J., Perdue, G. N., Pedro, K., ... Wild, S. M. (2022, September). DeepAdversaries: examining the robustness of deep learning models for galaxy morphology classification. *Machine Learning: Science and Technology, 3*(3), 035007. doi: 10.1088/2632-2153/ac7f1a

Cireşan, D., Meier, U., & Schmidhuber, J. (2012, February). Multi-column Deep Neural Networks for Image Classification. *arXiv e-prints*, arXiv:1202.2745. doi: 10.48550/arXiv.1202.2745

Connolly, A. J., Szalay, A. S., Bershady, M. A., Kinney, A. L., & Calzetti, D. (1995, September). Spectral Classification of Galaxies: an Orthogonal Approach. *AJ, 110*, 1071. doi: 10.1086/117587

Connor, L., Ravi, V., Catha, M., Chen, G., Faber, J. T., Lamb, J. W., ... Yadlapalli, N. (2023, February). Deep Synoptic Array science: Two fast radio burst sources in massive galaxy clusters. *arXiv e-prints*, arXiv:2302.14788. doi: 10.48550/arXiv.2302.14788

Connor, L., & van Leeuwen, J. (2018, December). Applying Deep Learning to Fast Radio Burst Classification. *AJ, 156*(6), 256. doi: 10.3847/1538-3881/aae649

Dai, Z., Moews, B., Vilalta, R., & Dave, R. (2023, March). Physics-informed neural networks in the recreation of hydrodynamic simulations from dark matter. *arXiv e-prints*, arXiv:2303.14090. doi: 10.48550/arXiv.2303.14090

DeLaunay, J., & Tohuvavohu, A. (2022, December). Harvesting BAT-GUANO with NITRATES (Non-Imaging Transient Reconstruction and Temporal Search): Detecting and Localizing the Faintest Gamma-Ray Bursts with a Likelihood Framework. *ApJ, 941*(2), 169. doi: 10.3847/1538-4357/ac9d38

Denker, J. S., & LeCun, Y. (1990). Transforming neural-net output levels to probability distributions. In *Advances in neural information processing systems 3* (p. 853–859). San Francisco, CA, USA: Morgan Kaufmann Publishers Inc.

Djorgovski, S. G., Baltay, C., Mahabal, A. A., Drake, A. J., Williams, R., Rabinowitz, D., ... Ellman, N. (2008, March). The Palomar-Quest digital synoptic sky survey. *Astronomische Nachrichten, 329*(3), 263. doi: 10.1002/asna.200710948

Drake, A. J., Djorgovski, S. G., Mahabal, A., Prieto, J. L., Beshore, E., Graham, M. J., ... Williams, R. (2012, April). The Catalina Real-time Transient Survey. In E. Griffin, R. Hanisch, & R. Seaman (Eds.), *New horizons in time domain astronomy* (Vol. 285, p. 306–308). doi: 10.1017/S1743921312000889

Dumusque, X. (2016, Aug). Radial velocity fitting challenge. I. Simulating the data set including realistic stellar radial-velocity signals. *A&A*, *593*, A5. doi: 10.1051/0004-6361/201628672

Dumusque, X., Borsa, F., Damasso, M., Díaz, R. F., Gregory, P. C., Hara, N. C., ... Udry, S. (2017, February). Radial-velocity fitting challenge. II. First results of the analysis of the data set. *A&A*, *598*, A133. doi: 10.1051/0004-6361/201628671

Dunn, M., Ćiprijanović, A., Nord, B., & Mobasher, B. (2023, January). Galaxy Morphology Classification Using Bayesian Neural Networks for LSST. In *American astronomical society meeting abstracts* (Vol. 55, p. 105.13).

Dvorkin, C., Mishra-Sharma, S., Nord, B., Villar, V. A., Avestruz, C., Bechtol, K., ... Villaescusa-Navarro, F. (2022, March). Machine Learning and Cosmology. *arXiv e-prints*, arXiv:2203.08056. doi: 10.48550/arXiv.2203.08056

Errington, T. M., Denis, A., Perfito, N., Iorns, E., & Nosek, B. A. (2021, dec). Reproducibility in cancer biology: Challenges for assessing replicability in preclinical cancer biology. *eLife*, *10*, e67995. Retrieved from `https://doi.org/10.7554/eLife.67995` doi: 10.7554/eLife.67995

Fukushima, K. (1980). Neocognitron: a self organizing neural network model for a mechanism of pattern recognition unaffected by shift in position. *Biol Cybern*, *36*(4), 193–202.

George, D., & Huerta, E. (2018a). Deep learning for real-time gravitational wave detection and parameter estimation: Results with advanced ligo data. *Physics Letters B*, *778*, 64–70. Retrieved from `https://www.sciencedirect.com/science/article/pii/S0370269317310390` doi: https://doi.org/10.1016/j.physletb.2017.12.053

George, D., & Huerta, E. A. (2018b, Feb). Deep neural networks to enable real-time multimessenger astrophysics. *Phys. Rev. D*, *97*, 044039. Retrieved from `https://link.aps.org/doi/10.1103/PhysRevD.97.044039` doi: 10.1103/PhysRevD.97.044039

Giorgi, G. M., & Gigliarano, C. (2017). The gini concentration index: A review of the inference literature. *Journal of Economic Surveys*, *31*(4), 1130–1148. doi: https://doi.org/10.1111/joes.12185

Graham, M. J., Djorgovski, S. G., Mahabal, A., Donalek, C., Drake, A., & Longo, G. (2012, August). Data challenges of time domain astronomy. *arXiv e-prints*, arXiv:1208.2480. doi: 10.48550/arXiv.1208.2480

Grojean, C., Paul, A., Qian, Z., & Strümke, I. (2022, March). Interpretable machine learning in Physics. *arXiv e-prints*, arXiv:2203.08021. doi: 10.48550/arXiv.2203.08021

Hartley, P., Bonaldi, A., Braun, R., Aditya, J. N. H. S., Aicardi, S., Alegre, L., ... Zuo, S. (2023, March). SKA Science Data Challenge 2: analysis and results. *arXiv e-prints*, arXiv:2303.07943. doi: 10.48550/arXiv.2303.07943

Hillar, C., & Sommer, F. (2012, October). Comment on the article "Distilling free-form natural laws from experimental data". *arXiv e-prints*, arXiv:1210.7273. doi: 10.48550/arXiv.1210.7273

Hložek, R., Ponder, K. A., Malz, A. I., Dai, M., Narayan, G., Ishida, E. E. O., ... Setzer, C. N. (2020, December). Results of the Photometric LSST Astronomical Time-series Classification Challenge (PLAsTiCC). *arXiv e-prints*, arXiv:2012.12392. doi: 10.48550/arXiv.2012.12392

Ho, J., Jain, A., & Abbeel, P. (2020, June). Denoising Diffusion Probabilistic Models. *arXiv e-prints*, arXiv:2006.11239. doi: 10.48550/arXiv.2006.11239

Hopkins, A. M., Whiting, M. T., Seymour, N., Chow, K. E., Norris, R. P., Bonavera, L., ... van der Horst, A. J. (2015, October). The ASKAP/EMU Source Finding Data Challenge. *Publ. Astron. Soc. Australia*, *32*, e037. doi: 10.1017/pasa.2015.37

Iyer, K. G., Speagle, J. S., Caplar, N., Forbes, J. C., Gawiser, E., Leja, J., & Tacchella, S. (2022, August). Stochastic Modelling of Star Formation Histories III. Constraints from Physically-Motivated Gaussian Processes. *arXiv e-prints*, arXiv:2208.05938. doi: 10.48550/arXiv.2208.05938

Jimenez, M., Alfaro, E. J., Torres Torres, M., & Triguero, I. (2023, February). CzSL: Learning from citizen science, experts and unlabelled data in astronomical image classification. *arXiv e-prints*, arXiv:2302.00366. doi: 10.48550/arXiv.2302.00366

Karniadakis, G. E., Kevrekidis, I. G., Lu, L., Perdikaris, P., Wang, S., & Yang, L. (2021, January). Physics-informed machine learning. *Nature Reviews Physics*, *3*(6), 422–440. doi: 10.1038/s42254-021-00314-5

Karpov, P. I., Huang, C., Sitdikov, I., Fryer, C. L., Woosley, S., & Pilania, G. (2022, November). Physics-informed Machine Learning for Modeling Turbulence in Supernovae. *ApJ*, *940*(1), 26. doi: 10.3847/1538-4357/ac88cc

Kessler, R., Bassett, B., Belov, P., Bhatnagar, V., Campbell, H., Conley, A., ... Varughese, M. (2010, December). Results from the Supernova Photometric Classification Challenge. *Publications of the Astronomical Society of the Pacific*, *122*, 1415. doi: 10.1086/657607

Kessler, R., Becker, A. C., Cinabro, D., Vanderplas, J., Frieman, J. A., Marriner, J., ... York, D. (2009, November). First-Year Sloan Digital Sky Survey-II Supernova Results: Hubble Diagram and Cosmological Parameters. *ApJS*, *185*(1), 32–84. doi: 10.1088/0067-0049/185/1/32

Kessler, R., Narayan, G., Avelino, A., Bachelet, E., Biswas, R., Brown, P. J., ... Transient and Variable Stars Science Collaboration (2019, Sep). Models and Simulations for the Photometric LSST Astronomical Time Series Classification Challenge (PLAsTiCC). *Publications of the Astronomical Society of the Pacific*, *131* (1003), 094501. doi: 10.1088/1538-3873/ab26f1

Kitching, T., Balan, S., Bernstein, G., Bethge, M., Bridle, S., Courbin, F., ... Voigt, L. (2010, September). Gravitational Lensing Accuracy Testing 2010 (GREAT10) Challenge Handbook. *arXiv e-prints*, arXiv:1009.0779. doi: 10.48550/arXiv.1009.0779

Klingner, C. M., Denker, M., Grün, S., Hanke, M., Oeltze-Jafra, S., Ohl, F. W., ... Ritter, P. (2022). Overcoming the reproducibility crisis - results of the first community survey of the german national research data infrastructure for neuroscience. *bioRxiv*. doi: 10.1101/2022.04.07.487439

Kruk, S., & Merín, B. (2023, April). Citizen Science with ESA Science Data - The Hubble Asteroid Hunter project. *Europhysics News*, *54*(2), 28–31. doi: 10.1051/epn/2023206

Lecar, M. (1968). Bulletin astronomique de l'observatoire de paris. In (Vol. 3). Centre National de la Research Scientifique.

Levrier, F., Wilman, R. J., Obreschkow, D., Kloeckner, H. R., Heywood, I. H., & Rawlings, S. (2009, January). Mapping the SKA Simulated Skies with the S3-Tools. In *Wide field astronomy & technology for the square kilometre array* (p. 5). doi: 10.22323/1.132.0005

Lewis, A., Voetberg, M., Nord, B., Jones, C., Hložek, R., Ciprijanovic, A., & Perdue, G. N. (2022, July). DeepBench: A library for simulating benchmark datasets for scientific analysis. In *Machine learning for astrophysics* (p. 32).

Lin, H.-H., Lin, K.-y., Li, C.-T., Tseng, Y.-H., Jiang, H., Wang, J.-H., ... Zhu, H.-M. (2022, September). BURSTT: Bustling Universe Radio Survey Telescope in Taiwan. *PASP*, *134*(1039), 094106. doi: 10.1088/1538-3873/ac8f71

Lintott, C. J., Schawinski, K., Slosar, A., Land, K., Bamford, S., Thomas, D., ... Vandenberg, J. (2008, September). Galaxy Zoo: morphologies derived from visual inspection of galaxies from the Sloan Digital Sky Survey. *MNRAS*, *389*(3), 1179–1189. doi: 10.1111/j.1365-2966.2008.13689.x

Lokken, M., Gagliano, A., Narayan, G., Hložek, R., Kessler, R., Crenshaw, J. F., ... LSST Dark Energy Science Collaboration (2023, April). The simulated catalogue of optical transients and correlated hosts (SCOTCH). *MNRAS*, *520*(2), 2887–2912. doi: 10.1093/mnras/stad302

Lorimer, D. R., Bailes, M., McLaughlin, M. A., Narkevic, D. J., & Crawford, F. (2007, November). A Bright Millisecond Radio Burst of Extragalactic Origin. *Science*, *318*(5851), 777. doi: 10.1126/science.1147532

LSST Dark Energy Science Collaboration. (2012, November). Large Synoptic Survey Telescope: Dark Energy Science Collaboration. *arXiv e-prints*, arXiv:1211.0310. doi: 10.48550/arXiv.1211.0310

LSST Dark Energy Science Collaboration (LSST DESC), Abolfathi, B., Alonso, D., Armstrong, R., Aubourg, É., Awan, H., ... Zuntz, J. (2021, March). The LSST DESC DC2 Simulated Sky Survey. *ApJS*, *253*(1), 31. doi: 10.3847/1538-4365/abd62c

LSST Science Collaboration, Abell, P. A., Allison, J., Anderson, S. F., Andrew, J. R., Angel, J. R. P., ... Zhan, H. (2009, December). LSST Science Book, Version 2.0. *arXiv e-prints*, arXiv:0912.0201. doi: 10.48550/arXiv.0912.0201

Lucie-Smith, L., Peiris, H. V., & Pontzen, A. (2023, May). Explaining dark matter halo density profiles with neural networks. *arXiv e-prints*, arXiv:2305.03077.

MacKay, D. J. C. (1992, 05). A Practical Bayesian Framework for Backpropagation Networks. *Neural Computation*, *4*(3), 448–472. doi: 10.1162/neco.1992.4.3.448

Malz, A., Hložek, R., Allam, J., Tarek, Bahmanyar, A., Biswas, R., Dai, M., ... Variable Stars Science Collaboration (2018, September). The Photometric LSST Astronomical Time-series Classification Challenge (PLAsTiCC): Selection of a performance metric for classification probabilities balancing diverse science goals. *ArXiv e-prints*, arXiv:1809.11145.

Mandelbaum, R., Rowe, B., Bosch, J., Chang, C., Courbin, F., Gill, M., ... Schrabback, T. (2014, May). The Third Gravitational Lensing Accuracy Testing (GREAT3) Challenge Handbook. *The Astrophysical Journal Supplement Series*, *212*, 5. doi: 10.1088/0067-0049/212/1/5

Megias Homar, G., Meyers, J. E., & Kahn, S. M. (2023, March). Prompt Detection of Fast Optical Bursts with the Vera C. Rubin Observatory. *arXiv e-prints*, arXiv:2303.02525. doi: 10.48550/arXiv.2303.02525

Metcalf, R. B., Meneghetti, M., Avestruz, C., Bellagamba, F., Bom, C. R., Bertin, E., ... Vernardos, G. (2018, February). The Strong Gravitational Lens Finding Challenge. *ArXiv e-prints*, arXiv:1802.03609.

Miller, R. H. (1964, July). Irreversibility in Small Stellar Dynamical Systems. *ApJ*, *140*, 250. doi: 10.1086/147911

Miller, R. H. (1971, January). Experimental studies of the numerical stability of the gravitational n-body problem. *Journal of Computational Physics*, *8*, 449–464. doi: 10.1016/0021-9991(71)90023-4

Modi, C., Lanusse, F., Seljak, U., Spergel, D. N., & Perreault-Levasseur, L. (2021, April). CosmicRIM: Reconstructing Early Universe by Combining Differentiable Simulations with Recurrent Inference Machines. *arXiv e-prints*, arXiv:2104.12864. doi: 10.48550/arXiv.2104.12864

Mohan, D., Scaife, A. M. M., Porter, F., Walmsley, M., & Bowles, M. (2022, April). Quantifying uncertainty in deep learning approaches to radio galaxy classification. *MNRAS*, *511*(3), 3722–3740. doi: 10.1093/mnras/stac223

Narayan, G., & ELAsTiCC Team. (2023, January). The Extended LSST Astronomical Time-series Classification Challenge (ELAsTiCC). In *American astronomical society meeting abstracts* (Vol. 55, p. 117.01).

Nord, B., Connolly, A. J., Kinney, J., Kubica, J., Narayan, G., Peek, J. E. G., ... Tollerud, E. J. (2019, September). Algorithms and Statistical Models for Scientific Discovery in the Petabyte Era. In *Bulletin of the american astronomical society* (Vol. 51, p. 224). doi: 10.48550/arXiv.1911.02479

Ntampaka, M., Ho, M., & Nord, B. (2021, November). Building Trustworthy Machine Learning Models for Astronomy. *arXiv e-prints*, arXiv:2111.14566. doi: 10.48550/arXiv.2111.14566

Ntampaka, M., & Vikhlinin, A. (2022, February). The Importance of Being Interpretable: Toward an Understandable Machine Learning Encoder for Galaxy Cluster Cosmology. *ApJ*, *926*(1), 45. doi: 10.3847/1538-4357/ac423e

Peek, J., & White, R. (2021, June). Search By Image: Citizen Science and Deep Learning for next-generation archives. In *American astronomical society meeting abstracts* (Vol. 53, p. 301.06).

Piras, D., Peiris, H. V., Pontzen, A., Lucie-Smith, L., Guo, N., & Nord, B. (2023, June). A robust estimator of mutual information for deep learning interpretability. *Machine Learning: Science and Technology*, *4*(2), 025006. doi: 10.1088/2632-2153/acc444

Psaros, A. F., Meng, X., Zou, Z., Guo, L., & Karniadakis, G. E. (2023, March). Uncertainty quantification in scientific machine learning: Methods, metrics, and comparisons. *Journal of Computational Physics*, *477*, 111902. doi: 10.1016/j.jcp.2022.111902

Quinlan, G. D., & Tremaine, S. (1992, December). On the reliability of gravitational N-body integrations. *MNRAS*, *259*(3), 505–518. doi: 10.1093/mnras/259.3.505

Raissi, M., Perdikaris, P., & Karniadakis, G. E. (2019, February). Physics-informed neural networks: A deep learning framework for solving forward and inverse problems involving nonlinear partial differential equations. *Journal of Computational Physics*, *378*, 686–707. doi: 10.1016/j.jcp.2018.10.045

Ramírez-Pérez, C., Sanchez, J., Alonso, D., & Font-Ribera, A. (2022, May). CoLoRe: fast cosmological realisations over large volumes with multiple tracers. *J. Cosmology Astropart. Phys.*, *2022*(5), 002. doi: 10.1088/1475-7516/2022/05/002

Razzano, M., Di Renzo, F., Fidecaro, F., Hemming, G., & Katsanevas, S. (2023, March). GWitchHunters: Machine learning and citizen science to improve the performance of gravitational wave detector. *Nuclear Instruments and Methods in Physics Research A*, *1048*, 167959. doi: 10.1016/j.nima.2022.167959

Reza, M., Zhang, Y., Nord, B., Poh, J., Ciprijanovic, A., & Strigari, L. (2022, July). Estimating Cosmological Constraints from Galaxy Cluster Abundance using Simulation-Based Inference. In *Machine learning for astrophysics* (p. 20). doi: 10.48550/arXiv.2208.00134

Ricker, G. R., Winn, J. N., Vanderspek, R., Latham, D. W., Bakos, G. Á., Bean, J. L., ... Villasenor, J. (2014, Aug). Transiting Exoplanet Survey Satellite (TESS). In *Space telescopes and instrumentation 2014: Optical, infrared, and millimeter wave* (Vol. 9143, p. 914320). doi: 10.1117/12.2063489

Riggi, S., Vitello, F., Becciani, U., Buemi, C., Bufano, F., Calanducci, A., ... Umana, G. (2019, October). Cuc(aesar) source finder: Recent developments and testing. *Publ. Astron. Soc. Australia*, *36*, e037. doi: 10.1017/pasa.2019.29

Rosofsky, S. G., Al Majed, H., & Huerta, E. A. (2022, March). Applications of physics informed neural operators. *arXiv e-prints*, arXiv:2203.12634. doi: 10.48550/arXiv.2203.12634

Rudin, C. (2018, November). Stop Explaining Black Box Machine Learning Models for High Stakes Decisions and Use Interpretable Models Instead. *arXiv e-prints*, arXiv:1811.10154. doi: 10.48550/arXiv.1811.10154

Sako, M., Bassett, B., Becker, A. C., Brown, P. J., Campbell, H., Wolf, R., ... Zheng, C. (2018, June). The Data Release of the Sloan Digital Sky Survey-II Supernova Survey. *PASP*, *130*(988), 064002. doi: 10.1088/1538-3873/aab4e0

Schmidt, M., & Lipson, H. (2009). Distilling free-form natural laws from experimental data. *Science*, *324*(5923), 81–85. doi: 10.1126/science.1165893

Selvaraju, R. R., Cogswell, M., Das, A., Vedantam, R., Parikh, D., & Batra, D. (2016, October). Grad-CAM: Visual Explanations from Deep Networks via Gradient-based Localization. *arXiv e-prints*, arXiv:1610.02391. doi: 10.48550/arXiv.1610.02391

Shukla, K., Xu, M., Trask, N., & Karniadakis, G. E. (2022, May). Scalable algorithms for physics-informed neural and graph networks. *arXiv e-prints*, arXiv:2205.08332. doi: 10.48550/arXiv.2205.08332

Siemiginowska, A., Eadie, G., Czekala, I., Feigelson, E., Ford, E. B., Kashyap, V., ... Young, C. A. (2019, May). The Next Decade of Astroinformatics and Astrostatistics. *BAAS*, *51*(3), 355. doi: 10.48550/arXiv.1903.06796

Square Kilometre Array Cosmology Science Working Group, Bacon, D. J., Battye, R. A., Bull, P., Camera, S., Ferreira, P. G., ... Zuntz, J. (2020, March). Cosmology with Phase 1 of the Square Kilometre Array Red Book 2018: Technical specifications and performance forecasts. *Publ. Astron. Soc. Australia*, *37*, e007. doi: 10.1017/pasa.2019.51

Stein, G., Seljak, U., Böhm, V., Aldering, G., Antilogus, P., Aragon, C., ... Nearby Supernova Factory (2022, August). A Probabilistic Autoencoder for Type Ia Supernova Spectral Time Series. *ApJ*, *935*(1), 5. doi: 10.3847/1538-4357/ac7c08

Sullivan, J. M., Prijon, T., & Seljak, U. (2023, March). Learning to Concentrate: Multi-tracer Forecasts on Local Primordial Non-Gaussianity with Machine-Learned Bias. *arXiv e-prints*, arXiv:2303.08901. doi: 10.48550/arXiv.2303.08901

The CHIME/FRB Collaboration, Andersen, B. C., Bandura, K., Bhardwaj, M., Boyle, P. J., ... Zwaniga, A. (2023, January). CHIME/FRB Discovery of 25 Repeating Fast Radio Burst Sources. *arXiv e-prints*, arXiv:2301.08762. doi: 10.48550/arXiv.2301.08762

Tohuvavohu, A., Kennea, J. A., DeLaunay, J., Palmer, D. M., Cenko, S. B., & Barthelmy, S. (2020, September). Gamma-Ray Urgent Archiver for Novel Opportunities (GUANO): Swift/BAT Event Data Dumps on Demand to Enable Sensitive Subthreshold GRB Searches. *ApJ*, *900*(1), 35. doi: 10.3847/1538-4357/aba94f

Tulio Ribeiro, M., Singh, S., & Guestrin, C. (2016, February). "Why Should I Trust You?": Explaining the Predictions of Any Classifier. *arXiv e-prints*, arXiv:1602.04938. doi: 10.48550/arXiv.1602.04938

Vafaei Sadr, A., Vos, E. E., Bassett, B. A., Hosenie, Z., Oozeer, N., & Lochner, M. (2019, April). DEEPSOURCE: point source detection using deep learning. *MNRAS*, *484*(2), 2793–2806. doi: 10.1093/mnras/stz131

van Leeuwen, J., Kooistra, E., Oostrum, L., Connor, L., Hargreaves, J. E., Maan, Y., ... Ziemke, J. (2023, April). The Apertif Radio Transient System (ARTS): Design, commissioning, data release, and detection of the first five fast radio bursts. *A&A*, *672*, A117. doi: 10.1051/0004-6361/202244107

van Roestel, J., Duev, D. A., Mahabal, A. A., Coughlin, M. W., Mróz, P., Burdge, K., ... Kulkarni, S. R. (2021, June). The ZTF Source Classification Project. I. Methods and Infrastructure. *AJ*, *161*(6), 267. doi: 10.3847/1538-3881/abe853

Vanderlinde, K., Liu, A., Gaensler, B., Bond, D., Hinshaw, G., Ng, C., ... Kaspi, V. (2019, October). The Canadian Hydrogen Observatory and Radio-transient Detector (CHORD). In *Canadian long range plan for astronomy and astrophysics white papers* (Vol. 2020, p. 28). doi: 10.5281/zenodo.3765414

Walmsley, M., Smith, L., Lintott, C., Gal, Y., Bamford, S., Dickinson, H., ... Wright, D. (2019, 10). Galaxy Zoo: probabilistic morphology through Bayesian CNNs and active learning. *Monthly Notices of the Royal Astronomical Society*, *491*(2), 1554–1574. Retrieved from `https://doi.org/10.1093/mnras/stz2816` doi: 10.1093/mnras/stz2816

Wang, Y.-T., Liu, H.-Y., & Piao, Y.-S. (2023, February). Self-supervised learning for gravitational wave signal identification. *arXiv e-prints*, arXiv:2302.00295. doi: 10.48550/arXiv.2302.00295

Yong, S. Y., Hobbs, G., Huynh, M. T., Rolland, V., Petersson, L., Norris, R. P., ... Zic, A. (2022, November). SPARKESX: Single-dish PARKES data sets for finding the uneXpected - a data challenge. *MNRAS*, *516*(4), 5832–5848. doi: 10.1093/mnras/stac2558

Yu, W., Richards, G., Buat, V., Brandt, W. N., Banerji, M., Ni, Q., ... Yang, J. (2022, July). *Lsstc agn data challenge 2021*. Zenodo. Retrieved from `https://doi.org/10.5281/zenodo.6878414` doi: 10.5281/zenodo.6878414

Zhang, C., Wang, C., Hobbs, G., Russell, C. J., Li, D., Zhang, S. B., ... Ren, Z. Y. (2020, October). Applying saliency-map analysis in searches for pulsars and fast radio bursts. *A&A*, *642*, A26. doi: 10.1051/0004-6361/201937234

Zhou, B., Khosla, A., Lapedriza, A., Oliva, A., & Torralba, A. (2015, December). Learning Deep Features for Discriminative Localization. *arXiv e-prints*, arXiv:1512.04150. doi: 10.48550/arXiv.1512.04150

Zhu-Ge, J.-M., Luo, J.-W., & Zhang, B. (2023, February). Machine learning classification of CHIME fast radio bursts - II. Unsupervised methods. *MNRAS*, *519*(2), 1823–1836. doi: 10.1093/mnras/stac3599

Zuntz, J., Lanusse, F., Malz, A. I., Wright, A. H., Slosar, A., Abolfathi, B., ... LSST Dark Energy Science Collaboration (2021, October). The LSST-DESC 3x2pt Tomography Optimization Challenge. *The Open Journal of Astrophysics*, *4*(1), 13. doi: 10.21105/astro.2108.13418

Machine Learning in Astronomy: Possibilities and Pitfalls
Proceedings IAU Symposium No. 368, 2025
J. McIver, A. Mahabal & C. Fluke, eds.
doi:10.1017/S1743921323003502

Deep Machine Learning in Cosmology: Evolution or Revolution?

Ofer Lahav

Department of Physics and Astronomy, University College London, London WC1E 6BT, UK
email: o.lahav@ucl.ac.uk

Abstract. Could Machine Learning (ML) make fundamental discoveries and tackle unsolved problems in Cosmology? Detailed observations of the present contents of the universe are consistent with the Cosmological Constant Λ & Cold Dark Matter model, subject to some unresolved inconsistencies ('tensions') among observations of the Hubble Constant and the clumpiness factor. To understand these issues further, large surveys of billions of galaxies and other probes require new statistical approaches. In recent years the power of ML, and in particular 'Deep Learning', has been demonstrated for object classification, photometric redshifts, anomaly detection, enhanced simulations, and inference of cosmological parameters. It is argued that the more traditional 'shallow learning' (i.e. with pre-processing feature extraction) is actually quite deep, as it brings in human knowledge, while 'deep learning' might be perceived as a black box, unless supplemented by explainability tools. The 'killer applications' of ML for Cosmology are still to come. New ways to train the next generation of scientists for the Data Intensive Science challenges ahead are also discussed. Finally, the chatbot ChatGPT is challenged to address the question posed in this article's title.†

Keywords. methods: data analysis – methods: statistical – techniques: galaxies

1. Introduction

The exponential growth of Astronomical data is remarkable: for example, from samples of thousands and tens of thousands of galaxies in the 1980s (CfA, IRAS) to millions of galaxies in the 1990s and early 2000s (2dF, SDSS, BOSS), to 300 million galaxies at present (DES), and to billions of galaxies in new surveys (Rubin-LSST, Euclid) which will start in 2023–24. The data acquisition rate, number of galaxies, cost and number of researchers are given in Figure 1 below for some of the surveys.

The other revolution, across the whole of science and technology, is in the developments of Artificial Intelligence (AI) and Machine Learning (ML). Terminology-wise, ML is a subset of AI, and Deep Learning (DL) is a new approach within ML (see Figure 2). AI &ML in Astronomy is a very busy field now, with many papers published weekly, beyond what can be shown in a short presentation. I am taking the liberty to present here work over the years with my PhD students and Post-docs, and to use these as examples for a general discussion of AI & ML in Astronomy and beyond. While the application of ML in Astronomy is very common these days, there are open issues to what extent they are 'black boxes', the connection to traditional statistics and how to implement these methods. For recent comprehensive reviews see Baron (2019) on applications of

† This article is based on presentations of the Royal Astronomical Society George Darwin Lecture (2020), a plenary talk at the IAU Symposium 368 held in South Korea (2022) and a talk at the workshop 'Unsolved Problems in Astrophysics and Cosmology' at the Hebrew University (2022) .

Survey	Data per night/day	Galaxies	Cost	Scientists
DES	1 TeraB	~300 Million (all observed)	~$70M	~400
DESI	40 GigaB	~35 Million (12M observed)	~$70M	~600
Rubin-LSST	15 TeraB	~Billions	~$1.0B	~1000
Euclid	850 GigaB	~Billions	~$1.5B	~1500
SKA	1 PetaB	~Billions	~$1.3B	~1000

Figure 1. Examples of galaxy surveys: data rates, number of galaxies, financial resources and human capital. DES observations are complete, DESI is collecting data, and Rubin-LSST, Euclid and SKA will be operating coming years.

ML to Astronomy, Carleo et al. (2019) on ML across the Physical Sciences, and Huertas-Company & Lanusse (2022), on DL for galaxy surveys, with applications for object classification, galaxy properties, discovery, emulation of simulations and inference.

The outline of this article is as follows. The status of the Λ & Cold Dark Matter (ΛCDM) model is reviewed in the context of present and future galaxy surveys, that will chart billions of galaxies. Examples of AI & ML in Cosmology and Astrophysics are presented: (i) Object classification; (ii) Photometric redshifts; and (iii) Dark matter mapping and (iv) Simulation-based inference of the masses of the Milky Way and Andromeda and of cosmological parameters. The training of the next generation of astronomers as data scientists and the outlook for AI&ML in Astronomy is reviewed, with the example of UCL's Centre for Doctoral Training (CDT) in Data Intensive Science (DIS).

2. Background: the status of the cosmological model

Recent years have witnessed success after success of Einstein's General Relativity: from the measurement of a Cosmological Constant Λ (or its variants) to the detection of Gravitational Waves (GW) moving at the speed of light (LIGO collaboration 2017a). In more detail, recent cosmological measurements strongly favour a 'concordance' model in which the present-time universe is flat and contains approximately 5% baryons, 25% cold dark matter and 70% dark energy, with a small contribution from massive neutrinos (e.g. Planck collaboration 2018; Dark Energy Survey collaboration 2022; Lahav & Liddle 2021 for a review). The modern observational evidence for dark energy is attributed to Supernovae Ia surveys in the late 1990s (Perlmutter et al. 1999; Riess et al. 1998), with subsequent support from other observational probes, although there was already growing evidence for it in the previous decade. The concept of dark energy is a variant on Einstein's cosmological constant, Λ, and the proposition for a Λ-like linear force can be traced back to Newton's Principia (e.g. Calder and Lahav 2008). This ΛCDM paradigm and its extensions pose fundamental questions about the origins of the universe. Commonly, dark energy is quantified by an equation of state parameter w, which is the ratio of pressure to density, $w = P/\rho$. The case $w = -1$ corresponds exactly to Einstein's Cosmological Constant in General Relativity. But w may vary with cosmic epoch, e.g. in the case of scalar fields.

Essentially w affects both the geometry of the universe and the growth rate of structures. These effects can be observed via a range of cosmological probes, including the

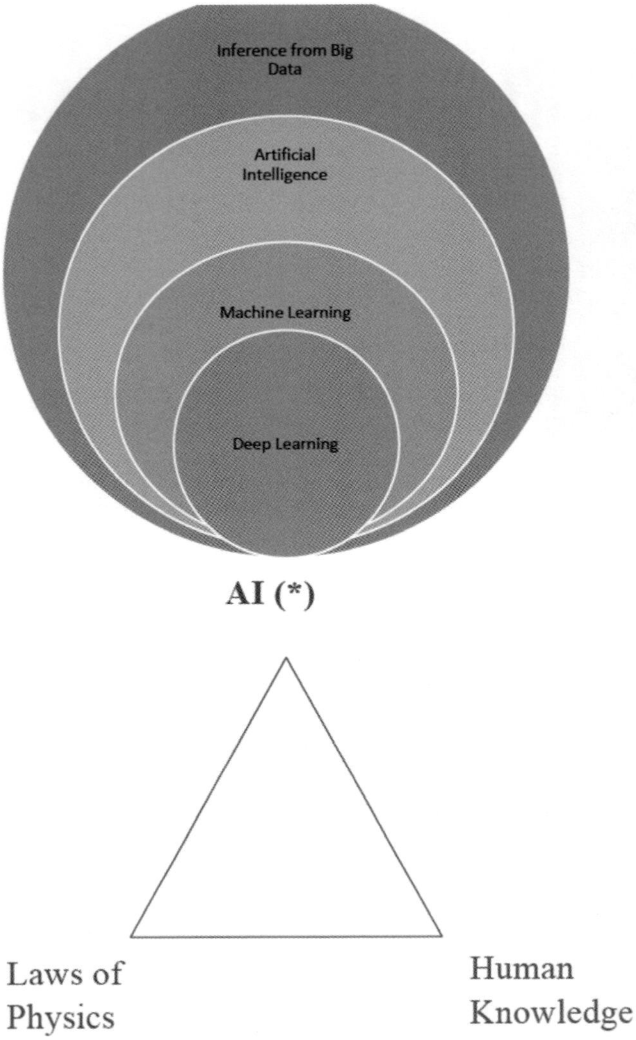

Figure 2. Top: The relationship between sub-fields: Statistical Inference, Artificial Intelligence, Machine Learning and Deep Learning, Bottom: the interplay between (1) 'objective' laws of Physics (2) current Human knowledge and (3) Augmented Intelligence that could update human knowledge based on new data.

Cosmic Microwave Background (CMB), galaxy clustering, clusters of galaxies, weak gravitational lensing and Supernovae Ia. For example, Supernovae Ia act as 'standard candles', while Baryonic Acoustic Oscillations (BAO) serve as 'standard rulers'. The current cosmological model, ΛCDM, has been spectacularly successful in fitting most of the observations. However, there is tension, at about 4σ, between the Hubble Constant $H_0 = 67.4 \pm 0.5$ km/sec/Mpc from CMB measurements (Planck collaboration 2018) and that measured locally $H_0 = 73.2 \pm 1.3$ km/sec/Mpc (Riess et al. 2021), that could force us to reconsider the standard ΛCDM model (for review see Shah et al. 2021). Unknown systematic errors might be the cause of this tension, and so we are in need of an independent probe of H_0 (e.g. gravitational wave 'bright' and 'dark' sirens) to confirm whether our model of the universe is in trouble. There is also tension, of about 2σ in the clumpiness parameter S_8, in the sense that the mass fluctuations measured by weak lensing in three different surveys seem a bit 'smoother' than what is measured by the Planck CMB.

Could Machine Learning help to tackle these problems in Cosmology? Some challenges and examples of ML applications are discussed below.

As an example of one of the surveys we highlight now the Dark Energy Survey (DES), in which I have been involved since 2004. DES completed in 2019 six observing seasons, over 758 nights. DES has used a wide-field camera on the 4m Blanco Telescope in Chile to image 5000 sq deg of the sky in five filters. The survey has generated a catalogue of 300 million galaxies with photometric redshifts and 100 million stars. In addition, a time-domain survey search over 27 sq deg has yielded a sample of thousands of Type Ia supernovae and other transients. The DES collaboration (2021) has measured $w = -1.03 \pm 0.03$ (68% CL) (using DES weak lensing & galaxy clustering combined with Planck and other probes, consistent with Einstein's Cosmological Constant. DES has also yielded unexpected science, including the discovery of distant objects in the Solar System, dwarf galaxies of the Milky Way, high-redshift Quasars and the detection of an optical flash from the LIGO Gravitational Wave Binary Neutron Star event GW170817. The story of DES, from an idea in 2003 through construction and observations to the first results in 2019 is described in the DES book (eds. Lahav et al. 2020).

The science of the next generation of surveys, 'Stage IV' (Rubin-LSST, DESI, Euclid, Roman-WFIRST) is even more ambitious. An intriguing question is "when to stop?" to measure cosmological parameters such as w (e.g. Lahav & Silk 2021). The short answer is that it depends what are the alternative theories. There is of course merit in increasing accuracy and precision of the measurements, which may reveal unexpected results. Detecting deviations from ΛCDM and measuring the neutrino mass would potentially be landmark measurements for the whole of Physics.

3. Deep vs. Shallow Learning

Astrophysics, High Energy Physics and essentially all fields in Science, Technology and Medicine are undergoing an 'industrial revolution', with huge data sets and complex models. For example, the above mentioned galaxy surveys will catalogue billions of galaxies over the coming decade. Similarly, Particle Physics experiments, e.g. at CERN, generate gigantic data sets. Such data sets call for an entirely new approach for data analysis. Fortunately, powerful methods have been invented by statisticians and computer scientists, and they are widely applied across different fields. In particular, ML techniques have gained a lot of popularity in recent years. Commonly used approaches are e.g. Artificial Neural Networks (ANN, see e.g. Figure 7), Boosted Decision Trees (BDT), and Support Vector Machines (SVM). See Various methods implemented e.g. in packages such as Scikit- Learn† and TensorFlow‡. A major breakthrough of recent years in ML is 'Deep Learning' (DL), see e.g. Le Cun et al. (2015) and Goodfellow et al. (2016). Rather than pre-processing to preform feature extraction first. For example, some DL algorithms implement a cascade of multiple layers ('filters'), which allow the full data to 'speak for themselves' (see e.g. Figure 5). DL can be used for both supervised and unsupervised learning. A popular implementation of DL is Convolutional Neural Networks (CNN). DL is certainly a major new trend in Astrophysics (Figure 3), with an increase from 23 papers with DL in their titles in 2017, to 105 in 2021. An example of a novel DL application is the detection of strong gravitational lensing (Metcalf et al. 2019), where CNN wins over visual inspection and other ML algorithms.

† `http://scikit-learn.org/stable/`
‡ `https://www.tensorflow.org/`

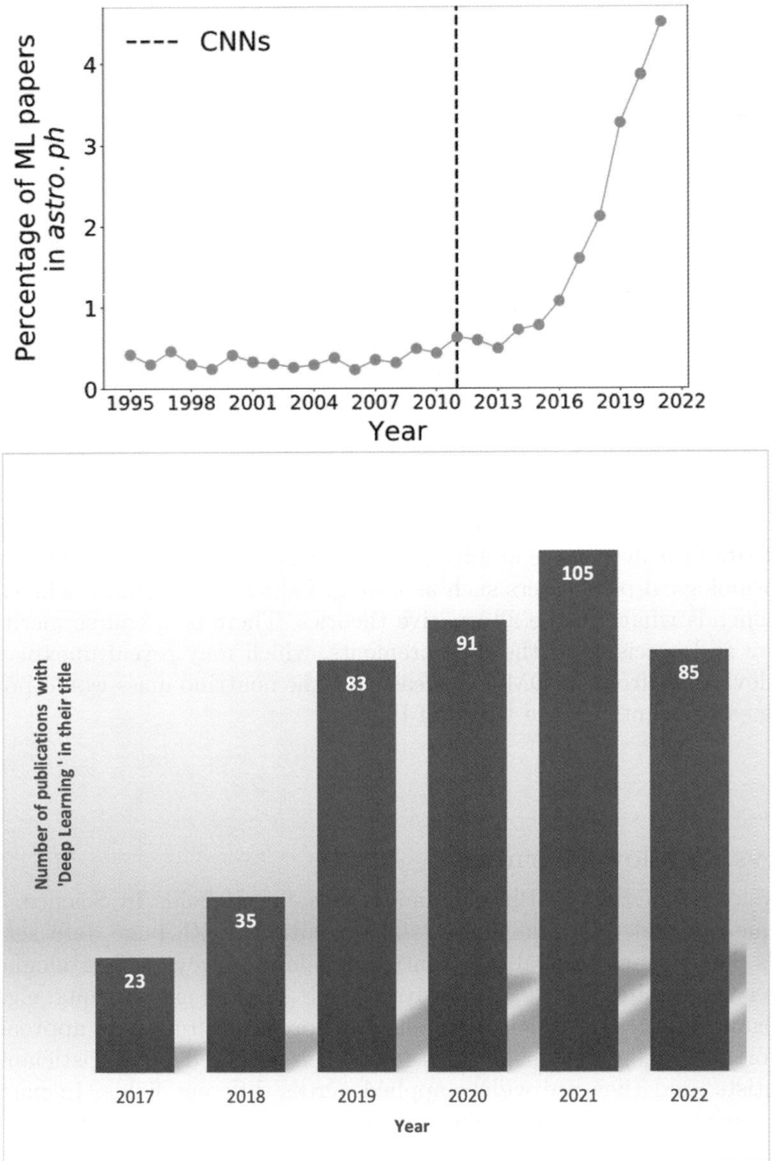

Figure 3. Top: The percentage of papers on astro-ph (arXiv.org) that report on at least one type of ML technique in the title or abstract. Some papers on ML for Astronomy appeared already prior to 1995 (see the text). The vertical line indicates the popularisation of Convolutional Neural Networks. Credit: D. Piras, PhD thesis (CDT-DIS/UCL). Bottom: The papers on astro-ph arXiv per year with 'Deep Learning' in their title.

While the performance of DL methods is impressive, it is still important to understand the DL algorithms, not to use them as just 'black boxes'. This can be done by running the ML algorithms through toy models, as well using *explainability* and *interpretability* tools (see below). We also argue that the traditional pre-processing of feature extraction in ML is not at all 'shallow'. A lot of human knowledge of Physics and Astronomy has influenced the choice of such features.

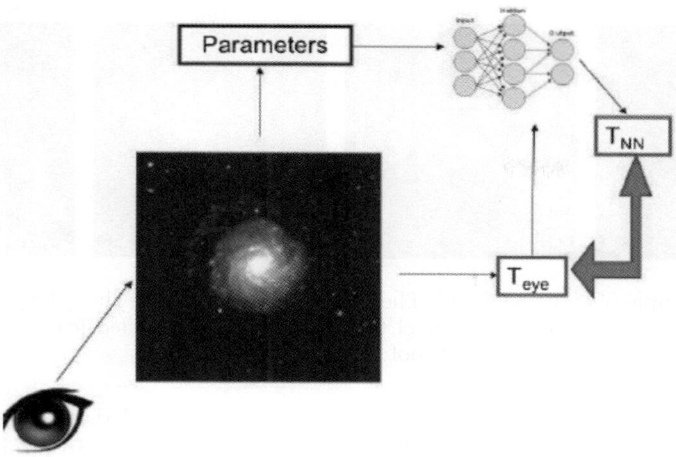

Figure 4. An illustration of supervised learning with Artificial Neural Networks for galaxy classification. The galaxy type classification of a training set is done by visual inspection of the entire image ($T_{\rm eye}$), while the training and testing only use a dozen or so features extracted by pre-processing from the galaxy images. The Neural Network weights are optimized to give $T_{\rm NN}$ as close as as possible to $T_{\rm eye}$, according to a metric defined by a cost function. The Graphics Credit: M. Banerji et al. (2010).

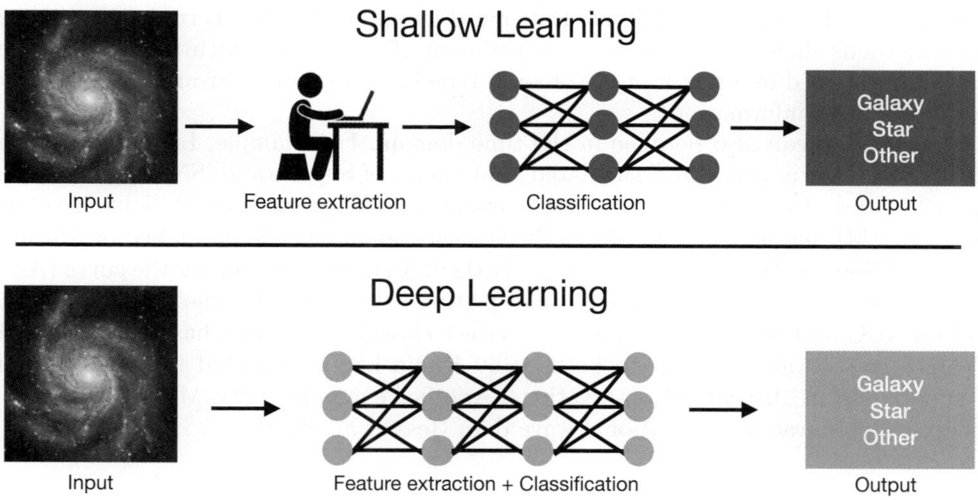

Figure 5. Contrasting 'Shallow Learning' (based on pre-processing feature extraction) with 'Deep Learning'. Graphics credit: S. M. Lahav, based on Wikipedia.

4. Object classification

Starting with a personal remark, I got interested in ML in the early 1990s, while at the Institute of Astronomy in Cambridge. I discussed with colleagues the possibility that computers, rather than humans, would classify galaxies. A chance encounter with a book about ANN was followed by work with Michael Storrie-Lombardi and Laerte Sodre, and later by the PhD work of Avi Naim (Storrie-Lombardi et al. 1992; Lahav et al. 1995, Naim et al. 1995). We defined an experiment of asking 6 classification 'gurus' to label 840 galaxy images (scanned by the Cambridge Automatic Plate Measuring machine). An ANN was then trained to use a dozen or so features to predict the de Vaucouleurs' T-system of 15 types. The ANN reproduced the classification to within 2 types. This work

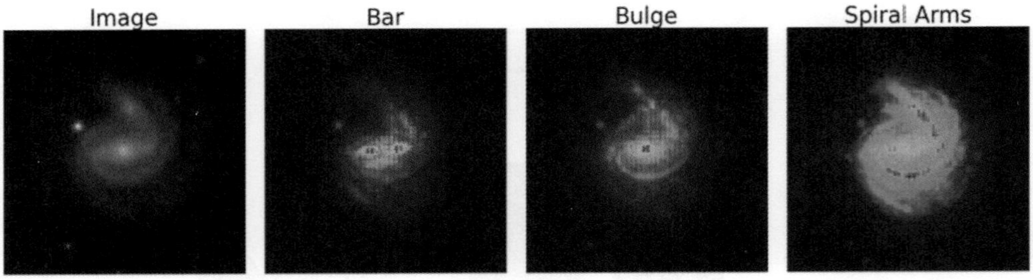

Figure 6. Example of explainable AI: The three saliency maps for the galaxy image on the left that indicate the sensitivity of each pixel to to the visually-classified features (bar, bulge and spiral arms), using an Explainable AI tool (Bhambra et al. 2022).

was outside the main-stream at the time, but was later extended for classification of the Galaxy Zoo (e.g. Banerji et al. 2010, Walmsley et al. 2020). It is an interesting question what features in the full image make the human decide on what type or another. A step towards this is by Explainable AI tools, such as SmoothGrad. Bhambra et al. (2022) have illustrated this by deriving saliency maps (see Figure 6), by calculating in each pixel the smoothed gradient of the visually-classified T-type with respect to the pixel intensity (bypassing the internal architecture of the algorithm).

Another important area is the classification of galaxy spectra. Examples include Principal Component Analysis (e.g. Madgwick et al. 2003) and information theory measures such as Entropy and Relative Entropy (Slonim et al. 2001; Ferreras et al. 2022). Most methods show that the information content of a spectrum with thousands of bins can be compressed by a factor of 100 or so. It remains to be seen if advanced ML methods can unveil more information.

Classification can also be done in the time domain. For example, Lochner, M., et al. (2016) have shown, using DES simulated light curves of Supernovae (SN) Type Ia how to extract features (using SALT2, parametric models and Wavelet) from SN light curves and to use five ML methods (Naïve Bayes, K Nearest-Neighbours, Support Vector Machine, ANN and Boosted Decision Trees) to classify them. Using the area under the curve (AUC) they have found for example that Wavelet feature extraction with the BDTs algorithm, give an AUC of 0.98, where 1 represents perfect classification. This has evolved into the SNMachine software package, to be used by Rubin-LSST, expected to detect millions of supernovae in 10 years of observations. Another example is the ML search for core collapse supernovae in gravitational wave data (Iess et al. 2022).

5. Photometric redshifts

Measuring the distance to galaxies is essential for cosmological studies such as galaxy clustering or weak gravitational lensing. Ideally, one would like the have a spectroscopic redshift for each galaxy. In reality, getting spectra requires long exposure times, so instead one can conduct multi-band photometry. Photometric redshifts ('photo-z') are not as precise as spectroscopic redshifts, but with many more galaxies with photometric redshifts than with spectroscopic redshifts it is possible to get reliable cosmological results. Photometric surveys such as DES, KiDS, HSC, Rubin-LSST, Euclid and Roman-WFIRST heavily rely on the photometric redshifts. It is probably fair to say that the deduction of photo-z is a pain for cosmologists (who would much prefer to use spectroscopic redshifts), but a joy for machine learners! Indeed, photo-z can be considered as an inverse problem, going from flux measured through several filters to the redshift (see Figure 7) and its associated probability distribution function. There are over a dozen

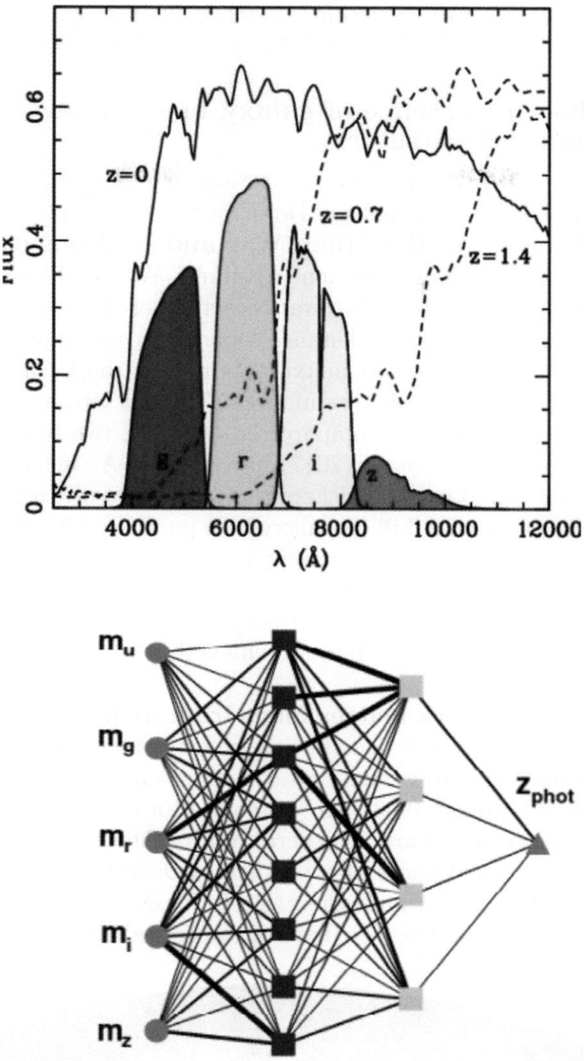

Figure 7. Top: A typical galaxy spectrum (flux against wavelength) at three different redshifts, as samples by four filters g,r,i,z. Graphics credit: Wikipedia. Bottom: An Artificial Neural Network configuration with 5 magnitudes (one per filter) and the redshift as output. Graphics credit: Sadeh et al. (2016).

methods in the literature for deriving photometric redshifts. Broadly speaking they are divided into template and training methods, where a small spectroscopic redshift data set can be used to train the ML tool. In our group we developed two decades ago an Artificial Neural Network software package starting from ANNz (Collister and Lahav 2004) and then ANNz2 (Sadeh et al. 2016). We've also attempted to improve the photo-z by adding galaxy structural parameters (Soo et al. 2018) and by a Deep Learning approach of analysing the full galaxy image (Henghes et al. 2022, following Pasquet et al. 2019). Furthermore, the photometric data can be used simultaneously to produce photo-z and stellar mass (Mucesh et al. 2021). We've also explored the scalability of ML algorithms for photo-z with the size of training data sets and other factors (Henghes et al. 2022). For other ML approaches to the photo-z problem, e.g. Self Organising Maps, see the review by Huertas-Company & Lanusse (2022). Photo-z with ML remain an active field for many more years, for Rubin-LSST, Euclid and other large photometric surveys.

6. Simulation-based inference of galaxy masses, dark matter maps and cosmological parameters

The advancements of both large simulations and ML algorithms have opened many new directions for research in Cosmology. Here are a few examples:

(i) The sum of masses of the Milky Way and Andromeda: This was initially estimated by the classical 'timing argument' (Kahn & Waltjer 1959; Lynden-Bell 1981), assuming infall of two point-like objects towards each other. The combined mass is derived based on the observed present-epoch distance separation, the infall velocity and the age of the universe. But in reality the two galaxies are part of the Local Group, hosting tens of galaxies, and itself residing in the cosmic web. To take into account this complexity of structure and dynamics one can compare the observed two-body system with many simulated galaxy pairs. McLeod et al. (2017) have trained Artificial Neural Networks on 30,000 simulated galaxy halo pairs, and Lemos et al. (2021) utilized Density Estimation Likelihood Free Inference with 2 million galaxy halo pairs, extracted from LCDM simulations. The latter study derived for the combined mass enclosed within density contrast of 200, $M_{200} = 4.6 \times 10^{12}\,{}^{+2.3}_{-1.8}$ (68% CL) solar masses. The estimated value is in accord with the timing argument, but somewhat larger than the sum of individual masses derived by other method. The error bars are much more refined due to the advanced ML approaches employed.

(ii) Reconstruction of dark matter maps from weak lensing: The inverse problem of deriving the projected mass distribution from the images of distant galaxy images distorted ('cosmic shear') due to weak gravitational lensing was addressed by the seminal algorithm of Kaiser & Squires (1993). An example of a reconstructed mass map (using a Wiener filter) is shown in Figure 8. Several approaches to this reconstruction problem have been taken in recent years, among them a Deep Learning approach technique (Jeffrey et al. 2021b). The algorithm learns the relationship between cosmic shear to projeted mass from about 10^5 simulations. Then the observed shear (e.g, 100 million galaxy

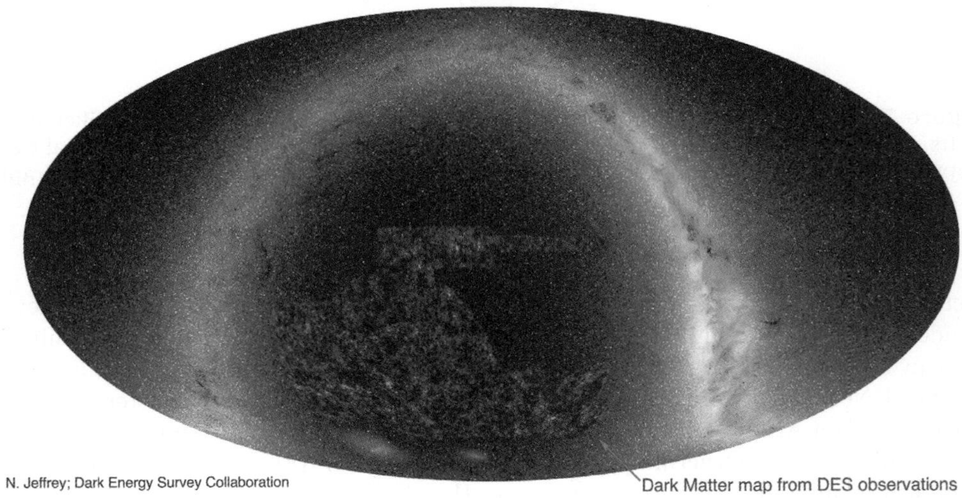

Figure 8. Mass map derived from DES Year 3 Weak Lensing, using 100 million galaxies with shape measurements over the 5000 survey area. High and low density regions are shown in bright and dark colours, respectively. The Milky Way image from Gaia is shown for the sky orientation. Credit: Nial Jeffrey & DES collaboration (2021).

shape measurements from DES over 5000 sq deg of the sky) is presented to the algorithm to produce a dark matter map. These techniques can be applied further to Euclid and Rubin-LSST (each of the two covering about 20000 sq. deg., with expected 1 billion galaxies). Such dark matter map are very useful for comparison with clusters and voids in galaxy maps (Hang et al., in preparation), and other multi-wavelength observations of the large scale structures. More generally, maps are very useful for studying structure at the 'field level', beyond the commonly used 2-point statistics. The galaxy and dark matter maps can be used for N-point statistics (e.g. Gualdi et al. 2019), probability distribution functions (e.g. Clerkin et al. 2017) and graph statistics such as Minimum Spanning Tree (Naidoo et al. 2020). On small scales, weak lensing maps can uncover the underlying dark matter distributions of halos and galaxy clusters. This can be used to study the effect of baryons and neutrinos on halo substructure, and particle interactions of dark matter on merging galaxy clusters. Another promising direction is to use DL to train on maps to predict cosmological parameters (Kacprzak and Fluri 2022), (Jeffrey et al. 2021a), and to even explain, via saliency maps (similar to Figure 6) the regions in the map that tell us most about certain cosmological and astrophysical parameters). It is important that the simulations are as realistic as possible and the systematics are modelled properly.

(iii) ML for gravitational physics: Another growing area is in analyses of N-body and hydrodynamic simulations. For example, Lucie-Smith et al. (2018) used ML to study the relationship between the initial conditions in simulations and the final dark haloes. ML can enhance large number of 'cheap' simulations, by training on a small number of 'expensive' simulations, e.g. the Lagrangian Deep Learning of Dai & Seljak (2021). An even more ambitious goal is to let an ML algorithm to discover new physics, for example re-discovering the orbital mechanics solar system Lemos et al. (2022).

7. Training the next generation of astronomers as data scientists

The fast-moving AI technology requires a new type of training, for a dual career in Astrophysics (High Energy Physics or other branches of Physics) as well as in Data Intensive Science (DIS). As an example, the UK's STFC has initiated the new scheme of Doctoral Training Centre (CDT) in DIS. Our CDT-DIS at UCL† trains about 10 CDT PhD students per year. This is a 4-yr PhD programme, including a 6-month placement in one of 30 or so private, public or third sector DIS organisations, e.g. from Natural Language Processing for fashion retail to ML for nuclear fusion. The PhD programme is now part of our UCL Centre for Data Intensive Science and Industry (DISI), which also include a one-year MSc programme for about 70 students. The Centre has strong links with international Astronomy and HEP projects and with other universities. With the support of the UK's Newton's Fund we also run a DIS training programme with Jordan‡.

8. Outlook

AI is transforming the world. We have already reached the stage where we cannot analyse cosmological experiments without ML of some sort. ML certainly enhances productivity. The relatively new development of Deep Learning has already proved very useful for problems in Astronomy such as object classification, photometric redshifts, anomaly detection, enhanced simulations, and inference of cosmological parameters. However, I argue that the more traditional 'shallow learning' (i.e. with pre-processing feature extraction) is actually quite deep, as it brings in human knowledge, while 'deep learning' might be perceived as a black box, unless supplemented by explainability tools. The relationship of AI, laws of physics and human knowledge is interconnected, with

† `https://www.ucl.ac.uk/data-intensive-science-cdt/`
‡ `https://www.ucl.ac.uk/astrophysics/research/ucl-jordan-dis-collaboration`

cross-fertilisation. As advocated by some, AI can be viewed as 'Augmented Intelligence', rather than as 'Artificial Intelligence'. In this context AI is enhancing human intelligence rather than replacing it (e.g. Lochner M. and Bassett B. A. 2021).

There are other challenges: Can we explain/interpret ML algorithms without using them just as 'black boxes'? How to minimise biases due to incomplete training sets? Can we include prior Physics? Can we learn new Physics? How to best train the next generation of astronomers also as data scientists? This takes us to the next level, applying novel AI and ML methods galaxy surveys (DES, DESI, Rubin-LSST, Euclid and others), with a huge discovery potential, from detecting subtle deviations in the standard LCDM model, measuring the neutrino mass and possibly discovering new populations of astronomical objects. It is probably fair to say that we are still missing 'killer applications' of ML in cosmology, something like the recent success of ML for Protein Folding†.

Finally, I have asked the AI chatbot ChatGPT‡ the question posed by this article's title: "Is Deep Machine Learning for Cosmology an evolution or a revolution?". The chatbot has replied:

"Deep machine learning has the potential to be both an evolution and a revolution in the field of cosmology. On one hand, it represents an evolution in the sense that it builds upon and extends existing methods and approaches that have been used in cosmology for decades. For example, machine learning algorithms can be used to analyse and interpret large datasets in ways that are more efficient and accurate than traditional methods.

On the other hand, deep machine learning also has the potential to be a revolution in cosmology by enabling the discovery of new patterns and insights that were previously hidden or difficult to discern. For example, machine learning algorithms can be used to identify complex correlations and patterns in cosmological data that may not be easily identifiable through traditional methods of analysis. In this way, deep machine learning has the potential to fundamentally change the way that cosmologists approach and study the universe."

Let's hope the ChatGPT's prediction in their last sentence is correct!

Acknowledgements

I thank my PhD students, Post-docs and collaborators over the years for their contribution to the work presented here. I acknowledge an Advanced ERC and STFC consolidated and CDT-DIS grants for their support of this research, and a Visiting Fellowship at All Souls College Oxford.

References

Banerji, M., et al., 2010, MNRAS, 406, 342
Baron, D., et al., 2019, arXiv:1904.07248
Bhambra, P., et al, 2022, MNRAS, 511, 5032
Bishop, C., 2006, *Pattern recognition and Machine Learning*, Springer
Calder, L. & Lahav, O., 2008, RAS Astronomy & Geophysics, 49, 1.13
Carleo, G., et al., 2019, Rev Mod Phys, 91, 045002
Chang, C., et al., 2016, MNRAS, 459, 3203
Clerkin, L., et al., 2017, MNRAS, 466, 1444
Collister, A. & Lahav, O., 2004, PASP, 116, 345
Dai, B. & Seljak, U., 2021, PNAS, 118 (16) e2020324118
Dark Energy Survey collaboration, 2016, MNRAS, 460, 1270
Dark Energy Survey collaboration, 2022, Phys Rev D, 105, 023520
Elgaroy, O,. et al., 2002, Phys Rev Lett, 89, 061301

† `https://www.deepmind.com/research/highlighted-research/alphafold`
‡ chat.openai.com

Ferreras, I., et al., 2022, RASTI (in press), arXiv:2208.05489
Goodfellow, I., et al. ., 2016, *Deep Learning*, MIT Press
Gualdi, D., et al., 2019, MNRAS, 484, 3713
Henghes, B., et al., 2021, MNRAS, 505, 4847
Henghes, B., et al., 2022, MNRAS, 512, 1696
Huertas-Company, M. & Lanusse, F., 2022, arXiv: 2210.01813
Iess, A., et al., 2022, A&A, 669, A42
Jeffrey, N., et al., 2021a, MNRAS, 501, 954
Jeffrey, N., et al., 2021b, MNRAS, 505, 4626
Kacprzak, T. & Fluri, J., 2022, arXiv:2203.096116
Kahn, F.D., & Woltjer, L., 1959, ApJ, 130, 705
Kaiser, N. & Squires G., 1993, ApJ, 404, 441
Lahav, O., et al., 1995, Science, 267, 859
Lahav, O. & Liddle, A., 2021, Reviews of Particle Physics, arXiv:2201.08666
Lahav, O. & Silk, J., 2021, Nature Astronomy, 5, 855
Lahav, O., et al. (eds.), 2020, *The Dark Energy Survey: The Story of a Cosmological Experiment*, World Scientific
Le Cun, Y., et al., 2015, Nature, 521, 43
Lemos, P., et al., 2021, Phys. Rev. D, 103, 023009
Lemos, P., et al., 2022, arXiv:2202.02306
LIGO collaboration, 2017a, ApJ, 848, L12
LIGO collaboration, 2017b, Nature, 551, 85
Lochner, M., et al., 2016, ApJS, 225, 31
Lochner, M. & Bassett, B.A., 2021, Astronomy and Computing, 36, 100481
Lucie-Smith, L., et al., 2018, MNRAS, 479, 3405
Lynden-Bell, D., 1981, Observatory, 101, 111
Madgwick, D.S., et al., 2003, MNRAS, 343, 871
McLeod, M., et al., 2017, JCAP, 12 03
Metcalf, R.B., et al., 2019, A&A, 625, A119
Mucesh, S., et al., 2021, MNRAS, 502, 2770
Naidoo, K., et al., 2020, MNARS, 491, 1709
Naim, A., et al. 1995, MNRAS, 275, 567
Pasquet, J., et al., 2019, A & A, 621, A26
Perlmutter, S., et al. 1999, ApJ, 517, 565
Planck collaboration, 2018, arXiv:1807.06209
Riess, A.. et al., 1998, AJ, 116, 1009
Riess, A., et al., 2021, ApJ, 853, 126
Sadeh, I., et al.. 2016, PASP, 128, 4502
Shah, P., et al., 2021, A&A Review, 29, 9
Slonim, N., et al. 2001, MNRAS, 323, 270
Soo, J., et al., 2018, MNRAS, 475, 3613
Storrie-Lombardi, M..C., et al., 1992, MNRAS, 259, 8p
Walmsley, M., et al., 2020, MNRAS, 491, 1554

Machine Learning in Astronomy: Possibilities and Pitfalls
Proceedings IAU Symposium No. 368, 2025
J. McIver, A. Mahabal & C. Fluke, eds.
doi:10.1017/S1743921323001710

An Astronomers Guide to Machine Learning

Sara A. Webb[1] and Simon R. Goode[1,2]

[1]Centre for Astrophysics and Supercomputing, Swinburne University of Technology, Hawthorn, Melbourne, Australia
email: swebb@swin.edu.au

[2]OzGrav ARC Centre of Excellence for Gravitational Wave discovery, Swinburne University of Technology, Hawthorn, Melbourne 3122, Australia

Abstract. With the volume and availability of astronomical data growing rapidly, astronomers will soon rely on the use of machine learning algorithms in their daily work. This proceeding aims to give an overview of what machine learning is and delve into the many different types of learning algorithms and examine two astronomical use cases. Machine learning has opened a world of possibilities for us astronomers working with large amounts of data, however if not careful, users can trip into common pitfalls. Here we'll focus on solving problems related to time-series light curve data and optical imaging data mainly from the Deeper, Wider, Faster Program (DWF). Alongside the written examples, online notebooks will be provided to demonstrate these different techniques. This guide aims to help you build a small toolkit of knowledge and tools to take back with you for use on your own future machine learning projects.

Keywords. Machine Learning, Observational, lightcurves

1. Introduction

In the field of artificial intelligence, machine learning focuses on using data and algorithms to mimic the way humans would typically learn, improving accuracy over time. Via machine learning, we can automate analytical models, taking advantage of algorithmic ability to learn from data and identify patterns with minimal human input.

Machine learning has already begun to be adopted by a wide range of sub-disciplines in astronomy, and is well established in some areas including it's use in transient astronomy as outlined in the advanced review by Fluke C. and Jacobs C. (2020). Note this work is not a comprehensive review of all techniques, rather a compilation of specific examples used within established transient astronomy programs.

In the era of large current and upcoming time-domain surveys, the classification and discovery of transient sources will rely on machine classification to handle large amounts of collected data. Current ground-based surveys such as the ZTF, DES and ASAS-SN scan thousands of square degrees per night, amounting to petabytes of data annually. Recently the Panoramic Survey Telescope and Rapid Response System Survey (Pan-STARRS) delivered the first-petabyte scale optical data release (Bellm et al. (2019); Amari et al. (2016); Shappee et al. (2014); Stubbs C. W., et al. (2010); Chambers K. C. et al. (2016)).

Space-based time-domain missions have provided unprecedented volumes of photometry, light curves, and proper motions for Galactic sources, with *Kepler* and K2 targeting ∼400,000+ individual stars, Transiting Exoplanet Survey Satellite (TESS, Ricker G. R. et al. (2014)) is expected to target at least 200,000 of the ∼9.5 million catalogue sources. The space-based mission, *Gaia*, is already releasing almost 2 billion

Supervised Learning			Unsupervised Learning	
Regression	Classification	Neural networks	Clustering	Dimensionality Reduction
Linear Regression	Logistic Regression	Convolutional Neural Networks (CNN)	K-means	t-Distributed Stochastic Neighbor Embedding (t-SNE)
K-Nearest Neighbors Regression (KNN)	K-Nearest Neighbors Classification (KNN)	Generative Adversarial Networks (GAN)	Gaussian Mixture Models (GMM)	Locally Linear Embedding (LLE)
Random Forest Regression	Random Forest Classification	Long Short Term Memory Networks (LSTM)	Hierarchical Agglomerative Clustering (HAC)	Uniform Manifold Approximation and Projection (UMAP)
Decision Tree Regression (CART)	Decision Tree classification (CART)	Gated Recurrent Units (GRU)	Density-Based spatial Clustering of Applications with Noise (DBSCAN)	Multidimensional Scaling (MDS)
Support Vector Regression (SVR)	Support Vector Machines (SVM)	Feedforward Neural Networks (FFNN)		Principal Component Analysis (PCA)
Multivariate Adaptive regression Splines (MARS)	Extreme Gradient Boosting (XGBoost)			Isomap Embedding
	Gradient Boosted Trees			
	Naive Bayes			

Figure 1. Representation of supervised verses unsupervised learning and the specific examples of each.

sources Borucki W. J., et al. (2010); Howell S. B., et al. (2014); Stassun K. G., et al. (2010).

Overcoming the mining challenges of these increasing amounts of data to not only identify and catalogue the multitude of known transient types but to discover additional new or anomalous sources is paramount to the success of future large transient surveys and time-domain science. This will become especially important with the upcoming Vera Rubin Observatory Legacy Survey of Space and Time (LSST, Abell P. A., et al. (2009)). LSST is a planned 10-year survey, imaging the entire Southern sky every three nights. LSST will generate millions of alerts each night, with billions of light curves continually updated or created.

Machine learning can be broken down broadly into either supervised or unsupervised algorithms, each pertaining subcategories beneath them. See Figure 1 for more details on the specific types of categories which will be explored below in more detail relating to the two real-data examples later on in the paper. For a comprehensive overview of techniques and uses in astronomy see Kembhavi A. and Pattnaik R (2022), Alzubi J. et al. (2018), and Mathew A. et al. (2020).

2. Overview

2.1. *Supervised Machine Learning*

One heavily used subcategory of machine learning in astronomy is supervised machine learning. This uses labelled data sets to train algorithms to classify data or predict outcomes. Traditionally a supervised algorithm would tune itself around the labelled data sets generating a function to map new inputs to likely outputs.

Supervised machine learning can be separated into three main branches: 1) Regression algorithms, 2) classification algorithms and 3) Neural networks. It is important to note that neural networks can be used to solved both regression and classification problems.

Regression algorithms focus around establishing the relationship between a single dependent variable that is dependent on several independent ones. Both linear and non-linear regression can be used for supervised learning. An example of a commonly used non-linear regression algorithm is that of Decision Trees (Quinlan K. G. (1986)). Decision Trees work by breaking down data into either decision or leaf nodes. Decision nodes are where the sub-node splits into further sub nodes, whereas the leaf nodes represent a final decision or terminal node. Once a decision tree is created on features from labeled data, the algorithm is able to assign a predicted value or outcome on subsequent data (Quinlan K. G. (1986)).

Classification algorithms are able to learn from a given labeled dataset to sort, assign and classify new data into a specific given number of classes or groups. Unlike regression algorithms which output continuous values, classification algorithms will only result in an assigned class or group. Classifications can be either binary or multi-class and are dependent on the problem being approached. A common classifier used in astrophysical contexts, including transient astronomy, is via a Decision Tree. Decision trees work by creating internal nodes and leaf nodes of grouping data around the features of the data. The internal nodes represent the conditions (e.g. certain features present or not) and the leaf nodes represent the decision made based on the conditions. These algorithms are useful for constructing easy to interpret tree like models around known data which can be used to classify new data.

As data is often plentiful in astronomy, supervised algorithms are ideal for completing the 'hard lifting' in classifying and sorting astronomical data sets.

One area which as heavily utilizes supervised algorithms is in the identification of variable sources. Variable stars and quasi-stellar objects have been identified from light curves via multivariate Gaussian mixture models, random forest classifiers, support vector machines, or Bayesian neural networks (Debosscher K. G. (2007); Richards J. W. et al. (2011); Kim D. et al. (2011); Pichara K. et al. (2012); Bloom J. S. et al. (2012); Pichara K. et al. (2013); Kim D. et al. (2016); Mackenzie C. et al. (2016); Muthukrishna, D. et al. (2022)).

All of the aforementioned work successfully uses classification of objects via supervised algorithms, which were trained on light curve extracted features. Features represent a set of measurable properties/characteristics of the light curves being studied. The most common features used in earlier works are available within the python package '*FATS*' by Nun I. et al. (2015).

Classification of non-folded light curves of extragalactic transient sources has also been explored, moving away from selecting the class of the object by fitting analytical templates built from a set of known sources Richards J. W. et al. (2011); Karpenka, N. V., et al. (2012); Lochner, M., et al. (2016); Narayan, G., et al. (2018); Moller, A., et al. (2016). While these techniques work well for catalogues of light curves, they cannot easily be applied to real-time data. Real-time classification of supernovae by Muthukrishna, D., et al. (2019) and Moller, A., et al. (2019) has shown the effectiveness of deep recurrent neural networks, without the need to rely on extracting computationally expensive features of the input data.

Another heavily explored use of supervised learning is for the determination or 'real' or 'bogus' sources in transient astronomy. These algorithms work by inspecting an image, or features of an image, and decide whether or not the image is of a genuine astrophysical source.

2.2. *Unsupervised Machine Learning*

With unsupervised machine learning, algorithms learn patterns from unlabeled data. These unsupervised algorithms self organise and create groupings or classes based on patterns exhibited as neuronal predilection or probability densities. These techniques are instrumental in finding like among like within a large data set.

Unsupervised machine learning can be separated into two main branches: 1) Clustering algorithms and 2) Dimensionality reduction algorithms.

Clustering algorithms work by grouping data into like clusters using features extracted from the data. The ultimate goal of clustering is to group like data together, and to isolate different clusters present within data. There are four main types of methods in clustering unlabelled data. The first is via density-based clustering, which works by identifying areas in feature space of high concentrations of data points. Distribution-based clustering takes the approach that all data points are part of the expected number of clusters, and works by calculating the probability that they belong to any given cluster. This works by using distance between points in feature space to determine the locations of clusters. Centroid-based clustering works by isolating the likely centroids within the data in feature space and determining relation to each cluster via distance metrics. Finally, hierarchical-based clustering works by organising data and groupings with a top–down approach, to insure groupings of varying densities are still identified as separate clusters.

In astronomy, clustering techniques have been used within recent large data sets to isolate known transient and variable sources using light curves (Valenzuela L. et al. (2018); Giles D. et al. (2019); Galarza M. et al. (2020)). This ability to cluster similar data will help identify previously unknown variables and transient events in the era of large astronomical surveys. As such, it will be invaluable to meaningfully and quickly quantify the expected large volume of short timescale events to help assist in follow-up priority assignment (Abell P. A., et al. (2009)).

Both have been explored within astronomical data and proved successful in providing meaningful insight into large data sets.

3. Examples of Applications in Astronomy

Here we outline and present resources to familiarize yourself with applying supervised and unsupervised machine learning to an observational astronomy data set. For both examples we'll be using different aspects of the Deeper Wider Faster (DWF) programs data.

The DWF program was developed to explore the fast dynamic universe, through multi-wavelength, multi-facility, real-time observations. The program is designed to run over ∼1 week observation blocks, at least once a year. The program is optimised to detect fast transients in real-time, and provide rapid follow-up with additional facilities. This first application of the programs brings up to the use of supervised machine learning in near-real time.

3.1. *Supervised Learning: Removal Of BOgus (ROBOT pipeline) for the Deeper Wider Faster program*

This section we outline one example of supervised learning from Goode S. et al. (2022), which can be further explored using the interactive notebooks on GitHub†.

During a typical DWF run, raw optical data (primarily from the CTIO Dark Energy Camera) is transferred in near real-time to Swinburne's OzSTAR supercomputer (Vohl, D., et al. (2017)). Each DECam image is comprised of 60 individual ccds, each 4K × 2K pixel resolution. The footprint of the imaging is large, covering ∼ 3 squared degrees.

Once the data has arrived on OzSTAR, it is processed for calibrations and made 'science ready' before being ingested through the *Mary* pipeline (Andreoni, I., et al. (2017)).

† https://github.com/simongoode/ROBOT-pipeline

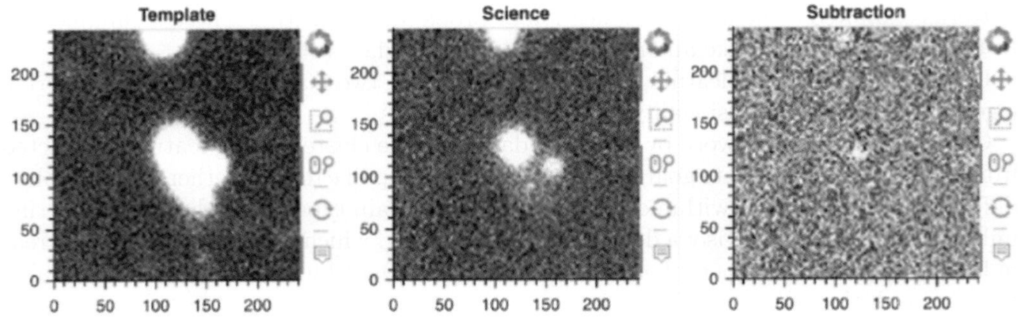

Figure 2. Example of real time processed candidate images, taken from a past operational run of DWF. Each panel is a small, 121 × 121 pixel image that corresponds to ∼ 30 × 30 arcsec on the sky centred on the candidate. The left panel is the (deeper) template image taken at a time previous to the DWF observations, the central panel is the current science image of the sky taken minutes earlier, and the subtraction image is the digital subtraction of the two images. All constant flux is subtracted, and any flux difference, e.g. from a transient source, will remain in the subtracted image.

Mary performs alignment and difference imaging between template and science images, and rapidly identifies transient candidates from positive residuals in the subtractions. During a single DWF observation run, hundreds of thousands of transient candidates are flagged through difference imaging, and it is crucial that promising candidates are inspected manually before triggering space and ground-based telescopes for follow-up observations. A large problem encountered during early DWF runs was the the immense data volume, which often exceeded the capability to all be accessed manually. With hundreds of thousands of candidate found in the processing, human inspectors were faced with more data then physically possible to evaluate without assistance. Astronomers were given several key pieces of information in a DWF run, one of which is the 'postage stamp' images of candidate sky location. Figure 2 shows an example of a transient candidate as processed in real time during a DWF run. Each of the three 'cutouts' is important for determining the realness of the the source, and what type of source it likely is.

The Removal Of BOgus Transients (ROBOT) pipeline was developed to significantly reduce the number of candidates needing human inspection, and rapidly improve the efficiency of candidate inspection during DWF observational runs (Goode S. et al. (2022)). The ROBOT pipeline aimed to work as an intermediary step between the processing/candidate identification and the image inspection preformed by astronomers. Due to the nature of the work, the uncertainties in sky conditions, and tendencies towards compression artifacts, a large majority of the data is often deemed as 'Bogus' or False positive alerts. ROBOT works to identify the astrophysical realness of objects, and filter only those of the highest likelihood through to the human inspector.

A deep Convolutional Neutral Network (CNN) was chosen to tackle this task, as historically CNNs have proven to be excellent for 2 dimensional data structures, like image data, and have a proven track record of reliable uses in image classification. The first step in building the ROBOT frame work was compiling and labeling large amounts of past DWF data into either real or bogus categories, and even specific source types.

Labelling of data occurred over several sessions by multiple expert astronomers. Using their past knowledge and experience, specifically around the DWF program, ensured contextually informed decisions. A total of 2952 candidate images containing template, science subtraction images were used. Out of the initial samples it was found 2250 were labelled unanimously by experts as bogus, and only 702 candidates were labelled as real. To limit the bias in the eventual network, we chose to balance the labelled data by

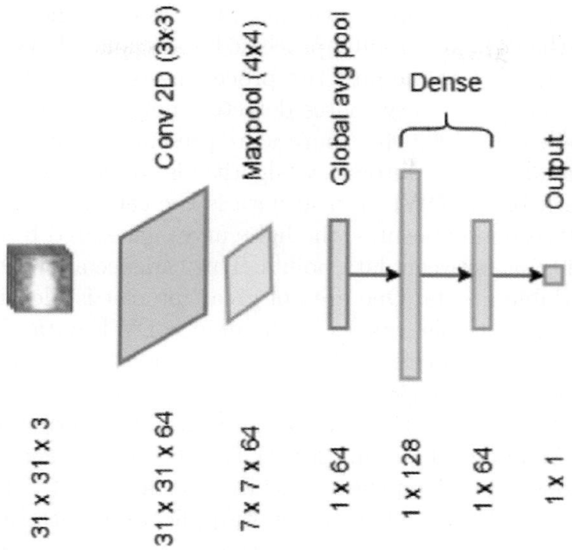

Figure 3. Figure from Goode S. et al. (2022). Architecture diagram of the highest performing model found during testing. The model takes in cropped triplet images (31 × 31 × 3) as input into a single 2D convolution layer, with 64 filters (3 × 3). A 4 × 4 maxpool layer passes 7 × 7 × 64 information to a global average pooling layer, which functions as both a dropout and flatten layer. This pooling layer passes the information to two fully connected dense layers, before providing a single output; the probability that the input triplet images belong to a real object.

using data augmentation. To do this, both the real and bogus images where included multiple times in the data set, but in different random augmentations including rotating, mirroring and translation. Each labelled set contained 5000 samples.

Using the labelled data, Goode S. et al. (2022) trained an initial 60 different CNN model architectures, each which slightly different combinations of layers, convolutions and hyperparameters. Each of the architectures was evaluated on their initial performance using the Matthews Correlation Coefficient, which took into account false positives, false negatives, true positives and true negatives to find a the architecture which preformed the best. It was found the best model was a '1c_2d' architecture, which consisted of 1 convolution, and 2 dense layers which are fully connected. The final architecture can be seen in Figure 3. The final algorithm preforms regression, returning an overall score between 0 and 1 for each the candidates passed through. The scores can then be used to determine the likelihood of it being either real or bogus, and can be used a classifier by setting the limits at which a final label of real or bogus is assigned to each.

By implementing ROBOT into DWF operations, the total time needed to inspect candidates was dramatically reduced, speeding up the human hours needed to make meaningful discoveries in real-time. Astronomy is uniquely positioned with the vast amounts of archival data able to be used in creating methods for automated discovery.

3.2. *Unsupervised Learning: Anomaly detection in lightcurves for the Deeper Wider Faster program*

This section we outline one example of unsupervised learning from Webb S. et al. (2020), which can be further explored using the interactive notebooks on GitHub†.

† https://github.com/sarawebb/ML_lightcurve_clustering

Although DWF is focused on chasing the fastest transient in near real-time during the observational runs, the data is still fully processed and explored systematically for other science objectives. One part of the post run processing is the production of light curves using the optical imaging, for every source detected, not just new transient sources. For a standard field, upwards of 100,000+ sources are present. To meaningfully evaluate this volume of sources, analytic and automated algorithms to identify sources of interest.

One exciting aspect of the DWF optical data is the cadence at which it is collected. Using continuous 20 second exposures, the lightcurves generated from this data have an average time of $\sim$ 60–75s between data points. This cadence allows high time resolution of transient and variable events. One area of great interest is identifying new or under explored transients and variable sources in the unique DWF data. To explored possibly unknown events we needed to design a flexible algorithm which could identify lightcurves which were anomalous to the majority.

We choose to use Hierarchical Density- Based Spatial Clustering of Applications with Noise (HDBSCAN, McLannes L. et al. (2017)).The theoretical method behind this algorithm was first proposed by Campello R. J. G. B., et al. (2013). HDBSCAN takes the approach of Density-Based Spatial Clustering of Applications with Noise (DBSCAN) and converts it into a hierarchical clustering algorithm by varying the value of epsilon (ϵ) to identify clusters of varying densities.

The power of HDBSCAN lies within it's ability to identify clusters of varying densities within a dataset. This is a valuable tool when working with diverse data such as lightcurves. Although we expect the majority of the sources to be unchanging over the short observational time periods, those which do change will do so in a variety of different ways, and not have consistent percentages of the data they occupy. Another advantage of HDBSCAN is it's ability to identify data which is highly anomalous and not belonging to any of the identified clusters. Before we are able to cluster the DWF lightcurves, we first needed to identify what features we wanted to use to describe each.

Features represent a set of measurable properties/characteristics of the light curves being studied. In this work we extracted a uniform set of features across all light curves for two purposes 1) reduce the dimensionality of the light curves and 2) allow for direct comparison between light curves that may be on different time scales with different sampling properties. We chose to use a mixture of normalised features developed and used primarily for the identification of variable stars and quasi-stellar objects, aiming to cluster variable and periodic sources, and have highly anomalous sources encapsulated in the unclusterable 'noise' unidentified by HDBSCAN. We began the feature selection by working on a sample sub sample of data, and calculating multiple different previously used, and data specific features. Using principle component analysis, we selected the top 25 with the largest eigenvalues. The chosen features are shown in Table 1 and explained in detail in Webb S. et al. (2020). We extracted these 25 unique features from each light curve using mostly using *FATS/FEETS* packages and some in-house routines (Nun I. et al. (2015)).

Using the lightcurves features we tested multiple configurations of HDBSCAN, changing minimum cluster size and distance metric type. After the preliminary tests we decided on the use of a minimum cluster size of 5 and the use of the Euclidean distance metric, for its intrinsic ability to calculate the shortest distance between points. These were chosen in an effort to create as many distinct clusters in our feature space as the algorithm will allow to limit the outliers to very low density regions.

In Webb S. et al. (2020) two separate fields/run types were explored 1) the DWF 'J04-55 field' which was data collected using a staring method on the telescope, and 2) 'Antlia field' collected using interleaved dithered. Both observational methods have their merits and uses, and we wanted to confirm that the clustering methods would work

Table 1. The details of each of the three clusters identified by the HDBSCAN algorithm. The description of the light curves refers to both the light curve and information gathered from individual cutouts of the detection images. unclustered represents light curves unable to be identified to a cluster.

Description of light curves	Cluster ID	# of Light Curves	% of Sources
Faint sources at detection threshold	Cluster 0	8	0.03%
Sources near CCD edge	Cluster 1	144	0.62%
Steady light curves	Cluster 2	22909	>98.7%
Real and photometrically affected light curves	Unclustered	138	< 0.59%

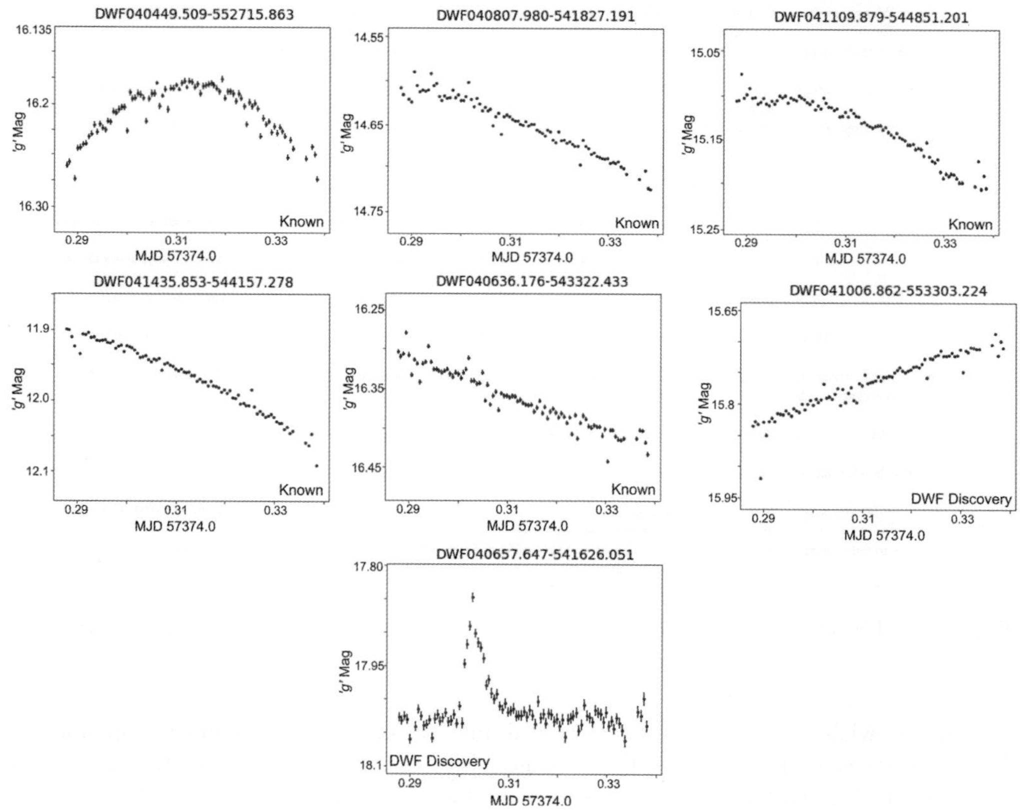

Figure 4. Four previously known and two newly discovered variable/transient sources present in the unclustered noise within the J04-55 field analysis.

to identify astrophysical variability as well as variability caused by observational effects such as dithering.

The clustering methods via HDBSCAN proved extremely successful in identifying not only distinct groupings of astrophysical sources, but also clustering lightcurves which were affected by observational effects such as dithering, blended sources or cosmic rays. Table 1 breaks down the cluster types identified using the DWF 'J04-55 field' of 23,199

Feature	Description	Inputs	Refs
Amplitudes	Half the difference between the median of the maximum 5% and the median of the minimum 5% Magnitude.	Magnitude	Richards et al. (2011)
Auto correlation length	Length of linear dependence of a signal with itself at two points in time	Magnitude	Kim et al. (2011)
Beyond1Std	Percentage of points beyond one standard deviation from the weighted mean	Magnitude & Error	Richards et al. (2011)
CAR_{mean}	The mean of a continuous time auto regressive model using a stochastic differential equation	Magnitude, Time & Error	Pichara et al. (2012)
CAR_{σ}	The variability of the time series on time scales shorter than τ	Magnitude, Time & Error	Pichara et al. (2012)
CAR_{τ}	The variability amplitude of the time series	Magnitude, Time & Error	Pichara et al. (2012)
H_1	Amplitude derived using the Fourier decomposition	Magnitude	Kim & Bailer-Jones (2016)
Con	The number of three consecutive data points that are brighter or fainter then 2σ and normalized by N -2	Magnitude	Kim et al. (2011)
Linear Trend	Slope of a linear fit to the light curve	Magnitude & Time	Richards et al. (2011)
MaxSlope	Maximum absolute magnitude slope between two consecutive observations	Magnitude & Time	Richards et al. (2011)
Mean	The mean magnitude	Magnitude	Kim et al. (2014)
Mean Variance	the ratio of the standard deviation to the mean magnitude	Magnitude	Kim et al. (2011)
Median Absolute Deviation	The median discrepancy of the data from the median data	Magnitude	Richards et al. (2011)
Median Buffer Range Percentage	Fraction of photometric points with amplitude/10 of the median magnitude	Magnitude	Richards et al. (2011)
Pair Slope Trend	The fraction of increasing first differences minus the fraction of decreasing first differences	Magnitude	Richards et al. (2011)
Q31	The difference between the 3rd and 1st quarterlies	Magnitude	Kim et al. (2014)
R_{21}	2nd to 1st amplitude ratio derived using the Fourier decomposition	Magnitude	Kim & Bailer-Jones (2016)
R_{31}	3rd to 1st amplitude ratio derived using the Fourier decomposition	Magnitude	Kim & Bailer-Jones (2016)
Rcs	Range of cumulative sum	Magnitude	Richards et al. (2011)
Skew	The skewness of the sample	Magnitude	Richards et al. (2011)
Slotted Auto Correlation Function Length	Slotted auto correlation length	Magnitude & Time	Protopapas et al. (2015)
Small Kurtosis	Small sample kurtosis of magnitudes	Magnitude	Richards et al. (2011)
Standard Deviation	Standard deviation of the magnitudes	Magnitude	Richards et al. (2011)
Stetson K_{AC}	Stetson K applied to the slotted auto correlation function of the light curve	Magnitude	Stetson (1996); Kim et al. (2011)
Variability Index	Ratio of the mean of the square of successive differences to the variance of data points	Magntiude	Kim et al. (2011)

Figure 5. Features used in Webb S. et al. (2020) and the properties of the light curves they represent.

lightcurves, with the distant clusters containing sources of unchanging magnitudes, or those at detection thresholds or CCDs edges. Interestingly in this field, the true astrophysical variable and transient sources were unable to be clustered and identified as noise. Figure 4 shows 7 such sources extracted from the grouping of noise.

For full analysis and the results from the 'Antlia field', including the use of *Astronomaly* and t-SNE's, and results see Webb S. et al. (2020) (see Figure 5).

4. Conclusions

Machine learning has already proven to be extremely powerful in it's ability to assist astronomers in discovery, and will only continue it's growth into more astronomical use cases. It is always important to note that machine learning isn't always a one solution fits all. It should be considered and applied with a great deal of care, to insure the problem tackled is solved in an efficient and unbiased manner. For those just beginning to explore the use of artificial intelligence in astronomical work, we highly recommend the use of existing frame works to evaluate the effectiveness of different methods. For the use of anomaly detection, the *Astronomaly* package is a flexible framework for use on both imaging and light curve data (Lochner M. and Bassett B. A. (2021)). It is undeniable that machine learning will shape the future of astronomy, with several large surveys already relying on intelligent algorithms.

References

Abell P. A., et al., 2010, *arXiv:0912.0201*

Alzubi J., et al., 2018, *J. Phys.: Conf. Ser.*, 1142, 012012

Abbott T., et al., 2 2016, *MNRAS*, 460, 1279

Andreoni, I., et al., 2017, *PASA* 34, E037

Bellm, E, et al., 2019, *PASP*, 131, 018002

Bloom J. S., et al. 2012, *PASP*, 124, 921

Borucki W. J., et al., 2010, *Science*, 327, 977

Campello R. J. G. B., et al., 2013, *Advances in Knowledge Discovery and Data Mining*, Springer Berlin Heidelberg, Berlin, Heidelberg, pp 160–172

Chambers K. C., et al., 2016, *arXiv:1612.05560*

Debosscher J., 2007, *A and A*, 475, 1159

Fluke C and Jacobs C. 2020, *WIREs Data Mining Knowl Discov.*,10:e1349

Galarza M., et al., 2020, *MNRAS*, 508, 4

Giles D., et al., 2019, *MNRAS*, 484, 1

Goode S., et al., 2022, *MNRAS*, 513, 2

Howell S. B., et al., 2014, *PASP*, 126, 398

Karpenka, N. V., et al., 2012, *MNRAS* 429, 2

Kembhavi A. and Pattnaik. R., et al., 2020, *J. Astrophys. Astron.*, 43, 76

Kim D., et al., 2016, *A and A*, 587, A18

Kim D., et al., 2011, *MNRAS*, 735, 2

Lochner, M., et al., 2016, *APJS* 225, 2

Lochner M. and Bassett. B. A., 2021, *Astronomy and Computing*, 36, 100481

Mackenzie C., et al., 2016, *APJ*, 820, 2

Mahabal A., et al., 2019, *PASP*, 131, 997

Mathew A., et al., 2020, *Adv. Intell. Syst.*, 1141

McLannes L., et al., 2017, *Astel*, 2, 11

Moller, A., et al., 2016, *JCAP* 2016,

Moller, A., et al., 2019, *MNRAS* 491, 3

Muthukrishna, D., et al., 2022, *MNRAS* 517, 1

Muthukrishna, D., et al., 2019, *PASP Special Issue on Methods for Time-Domain Astrophysics* 2016

Narayan, G., et al., 2018, *AAS* 236, 1

Nun I., et al., 2015, *arXiv:1506.00010*

Pichara K., et al. 2012, *MNRAS*, 427, 2

Pichara K., et al., 2013, *APJ*, 777, 2

Quinlan K. G., 1986, *Machine Learning*, 1, 81–106

Ricker G. R., et al., 2014, *SPIE*, 9143, 914320

Richards J. W., et al., 2011, *MNRAS*, 419, 1121

Roestel J., et al., 2016, *Astronomical Journal*, 161, 6

Shappee B. J., et al., 2014, *APJ*, 788, 48

Stassun K. G., et al., 2010, *AJ*, 156, 102

Stubbs C. W., et al., 2010, *APJ supplement Series*, 191, 376

Valenzuela L., et al., 2018, *MNRAS*, 474, 3

Vohl, D., et al., 2017, *PASA* 34, E038

Webb S., et al., 2020, *MNRAS*, 498, 3

Machine Learning in Astronomy: Possibilities and Pitfalls
Proceedings IAU Symposium No. 368, 2025
J. McIver, A. Mahabal & C. Fluke, eds.
doi:10.1017/S1743921322003647

Panel Discussion: Practical Problem Solving for Machine Learning

Guillermo Cabrera[1*], Sungwook E. Hong[2*], Lilianne Nakazono[3*], David Parkinson[2*] and Yuan-Sen Ting[4,5*]

[1]Department of Computer Science, University of Concepción, Chile
email: guillecabrera@inf.udec.cl

[2]Center for Theoretical Astronomy, Korea Astronomy and Space Science Institute, 776 Daedeok-daero, Yuseong-gu, Daejeon 34055, Republic of Korea
emails: swhong@kasi.re.kr, davidparkinson@kasi.re.kr

[3]Instituto de Astronomia, Geofísica e Ciências Atmosféricas da U. de São Paulo Cidade Universitária, 05508-900 São Paulo, SP, Brazil
email: lilianne.nakazono@usp.br

[4]Research School of Astronomy & Astrophysics, Australian National University, Cotter Road, Weston, ACT 2611, Australia
email: yuan-sen.ting@anu.edu.au

[5]School of Computing, Australian National University, Acton ACT 2601, Australia

[*]Equal contribution

Abstract. Machine Learning is a powerful tool for astrophysicists, which has already had significant uptake in the community. But there remain some barriers to entry, relating to proper understanding, the difficulty of interpretability, and the lack of cohesive training. In this discussion session we addressed some of these questions, and suggest how the field may move forward.

Keywords. Machine Learning, Astrostatistics, education, data science

1. Introduction

The practical utility of machine learning (ML) methods to astrophysical and astronomical problems has already been demonstrated for a multitude of different problems. Yet for many research astronomers, the mechanics of the ML toolbox remains somewhat mysterious, which leads some to use the methods without full understanding, and others to shun ML entirely. The path to adopting ML in astrophysics should be made clearer and easier, but without risking an over-simplification. In this discussion session we address some of the issues of machine learning interpretability, accessibility, and education.

2. Discussion Questions

2.1. *What is Interpretability in Astronomical ML Problems?*

SUNGWOOK: We astronomers are automatically astrophysicists. We want to *understand* the nature behind astronomical data, not just handle the data. But most ML techniques that guarantee high predictability are very sophisticated, and so become black boxes for most users. This means that, ironically, we now need to put additional efforts into *understanding* the very tool that we introduced to *understand* the nature. That is probably one of the key factors why some notable astronomers are disinclined towards applying MLs to astronomy.

While many so-called interpretable MLs have been suggested in the ML community, I believe that we need to discuss with classical, old-fashioned astronomers to define the *interpretability* in astronomical ML. Their expertise is *interpreting* the astronomical data and theories — their interpretations may not always be valid, but they have a very high probability of being valid. Therefore, if our definition and application of interpretability do not match with their intuition at all, it will not be accepted in the general astronomical society, no matter how fancy it seems to be.

YUAN-SEN: While interpretability is important for physical sciences, it also comes with many flavours and can vary drastically depending on the context. Take the example of N-body simulations. Complex systems with many "bodies" are fundamentally top-down processes. In such cases, it would be hard to argue that we could achieve analytic (bottom-up) interpretability. Despite that, I would argue that in many such systems, we can still understand the collective behaviour intuitively without chasing for the illusive bottom-up interpretability. I am often concerned that, in astronomy, we seem to hold a higher standard in terms of interpretability for machine learning than for, say, N-body simulations. I often think of deep learning as more like statistical mechanics. While it might not be possible to disentangle everything down to individual neurons, the collective behaviour of neural networks can be understood as intuitive. Its capability traces back to all the inductive biases rooted in various physical symmetries. And for that, I feel more optimistic about the interpretability issue. I would argue that, at the higher level, we have a decent mathematical understanding of deep learning on its collective or emergent behaviour. And this mathematical foundation of machine learning is bound to be a growing field in the next few years.

LILIANNE: There is a growing field of research in ML called Explainable AI (XAI) and one can find many systematic reviews in the literature. I particularly recommend start by reading Gilpin et al. (2018) They state:

> "The goal of interpretability is to describe the internals of a system in a way that is understandable to humans. [...] The challenge facing explainable AI is in creating explanations that are both complete and interpretable: it is difficult to achieve interpretability and completeness simultaneously. The most accurate explanations are not easily interpretable to people; and conversely the most interpretable descriptions often do not provide predictive power."

In simple terms, an explanation is an answer to a "why" question, i.e. "why a model predicts in a certain way?". The big "why" questions of the Universe are what inspire us as astronomers. Ideally, our goal should be to try to understand the relations between e.g. the input-output of a ML model, while also connecting them with theoretical physics. This is not an easy task, especially with the interpretability-completeness trade-off.

2.2. *Modelling versus Learning*

YUAN-SEN: A somewhat misconception about deep learning is that it is a total black box. On the one hand, as discussed above, the success of most, if not all, deep learning architectures can be traced back to their inductive biases and can be understood via their underlying physical symmetries. In this sense, deep learning itself is a type of modelling. On the other hand, any fitting even linear regression, is a type of learning. Modelling any physical system is always a delicate balancing act between modelling and learning and is never one or the other. As we impose fewer and fewer inductive biases (learning toward the learning side), as the solution space becomes larger, it will lead to a system that is more "training-data-hungry" and vice versa.

Viewed through this lens, often, the decider lies in understanding (a) how much knowledge we have on the fundamentals that drive the system and (b) how much data we have. If we have a profound understanding of the system (e.g. the exact equations that describe the system), modelling is favoured because it is less data-hungry. But quite often, we only know the symmetries (or even just approximate symmetries) that the system follows. In this case, perhaps the best we could do is to have a deep learning architecture that satisfies the symmetries. I would argue that this is not a more morally inferior position because choosing such modelling (i.e., the deep learning architecture) reflects our best understanding of the physical systems (the symmetries the systems have). What is more dangerous, however, is a blind application of any deep learning architecture without understanding the inductive biases that the architecture entails. In this case, the deep learning method will starve for data while at the same time subjecting us to the wrong solution space.

2.3. *Classical Machine Learning versus Deep Learning. When and why.*

LILIANNE: Following the discussion about interpretability-completeness tradeoff in 2.1, if one is trying to make interpretations from the data at some level, deep learning will be the last thing one would try out. Moreover, not all data and not all problem require complex solutions such as deep learning, so do not blindly go for deep learning (or even machine learning) without clearly understanding your problem. Now suppose one is not interested in interpretation but only cares about achieving accurate predictions. I would go for classical machine learning instead of deep learning in those scenarios: (1) when one lacks resources (e.g. GPU, cloud access), as deep learning requires more computational power by its nature; (2) when one has small data sets, as deep learning can easily overfit on them; (3) if one has no previous experience with deep learning or machine learning whatsoever (especially when time is tight!), as building a deep learning architecture can be very tricky, while classical machine learning are often more intuitive and easier to implement. Finally, in my point of view, deciding between deep learning and machine learning is also about time and resource management, balancing them with the gains that a deep learning model would provide. For example, if I have to spend 5x more time to train a deep learning model and to put it in production to only improve 0.01% in accuracy, it is probably not worth it.

YUAN-SEN: The key distinction between deep learning (i.e., neural network-based) and classical machine learning (regression, PCA, etc.) is that, while not fully understood, deep learning seems to evade the curse of dimensionality. And this is truly profound. As discussed Sec. 2.2, a critical goal of many physical modelling is to admit a solution space that is as generic as possible, yet the dimension does not curse it. Briefly, as the dimension increases, it does not require an exponentially large training data set to bound the fitting error. As we have grown to have a much better understanding of the mathematical foundation of deep learning, what has become more apparent is that the size of the solution space and the curse of dimensionality are not mutually exclusive. The fortuitous empirical success of neural networks has demonstrated that, with proper inductive biases, it is possible to admit a valuable ample solution space that is trainable with finite training data. Being able to perform a high-dimensional regression (supervised learning) or describe a high-dimensional probability density function (unsupervised learning) is a game changer that opens up many new possibilities for physical sciences, as most phenomenology in physical sciences is of inherently high dimension.

2.4. *Tailored Loss Functions for Specific Problems*

SUNGWOOK: As one who prefers applying DL to regression problems, most astronomical DL works seem to focus on enhancing architectures. However, once the architecture is complex enough, it is hard to achieve significant improvement by fine-tuning architecture or optimization methods.

Instead, I believe we need to focus more onto designing specific loss functions (and/or the normalization method of input and output dataset) optimized for the given problem. For example, suppose that one reconstructed the matter density using the mean-squared error (MSE) loss function. Then the machine will try to match the truth and prediction values at each position, but it does not guarantee that the reconstructed density field will have a similar large-scale feature to the ground truth. Some loss functions contain terms about large-scale features (Gatys, Ecker & Bethge 2016, Liu et al. 2018), but they do not guarantee to match the specific large-scale features in astronomical contents, such as power spectra or N-point correlation functions.

Another example is the intrinsic bias in the truth sample distribution. For example, in most region, the 3D dark matter density and the projected 2D dark matter density are around average or somewhat underdense (Hong et al. 2021a, Hong et al. 2021b). On the other hand, most information that astronomers want to get come from highly overdense regions, but using usual MSE-like loss functions does not guarantee excellent prediction at such overdense regions. To overcome this, one should manually emphasize important but less probable regions, such as by normalizing input or output density, adding customized weights to the loss function, etc (Lin et al. 2017). In conclusion, I believe astronomers need to focus more on finding loss functions optimized for what they want to match, instead of just adopting well-known functions in the ML community.

YUAN-SEN: Besides embedding inductive biases (physical symmetries) through the deep learning architecture, another way that has proven effective is to include our understanding of the systems, e.g., equations that the systems satisfy, directly in the loss function. Since we often have a lot more prior knowledge in physical sciences (e.g. PDE that the system follows), and some of these cannot easily materialise at the architecture level, tailoring the loss function to reflect this prior knowledge plays an exceedingly critical role in AI x Science.

2.5. *What is Necessary for Enhancing Astronomical ML Community?*

YUAN-SEN: To advance AI x Science, we need to cultivate a community that has a deep appreciation of both domains instead of treating machine learning as a tool that can be applied to astronomy. Deep learning (as opposed to classical machine learning) is a drastic paradigm shift in our understanding of mathematics (e.g., the curse of dimensionality). If the AI x Science community could act collectively to ensure other astronomers also see the true value and excitement in deep learning instead of just portraying deep learning as a tool, I believe it would lead to a positive perception change in the field. This resembles what the computational astronomy (hydrodynamical simulations) community has gone through in many ways. Yes, simulation techniques are useful for science, but at the same time, they are also profound, interesting mathematical objects. Realising the latter has sustained and grown the computational astronomy community drastically, attracting people interested in such mathematics beyond treating it as just a tool.

LILIANNE: Interdisciplinary is the keyword. Astronomical data sets are an immense (yet underutilised) playground for many theoretical researchers in Computer Science,

Statistics, and other STEM areas. Many of the problems we face with astronomical data (e.g. dealing with measurement errors, sample bias, etc) can serve as motivation for developing new techniques to address real-world problems. Therefore, if we want to enhance astronomical ML community, we must strengthen collaborations with other STEM fields. Building bridges between different fields are not easy, requiring an effort from all parts involved. And, of course, it is also important to encourage astronomy students who lean towards interdisciplinary. Being a graduate student myself and having my own concerns about my future in academia, I bring here some questions to think about: Is the academic system ready to really embrace cross-disciplinary people? Are evaluation metrics for academic productivity and tenure-tracked positions biased towards another profile? What cultural/systematic changes do we need in our departments, to start with?

2.6. *How to effectively train students in ML?*

SUNGWOOK: For me, the best way to learn astronomical ML is to (1) apply well-established ML techniques to astronomical problems, not the tutorial problems famous in the ML community, (2) see how they *fail*, and (3) find out how to improve it. When I first learned CNN, I was surprised at how easy its Python manifest is and how well such simple code can predict the number so well in the MNIST dataset. When I applied the same code to my problem later, I was shocked that it just did not work. But, this made me start thinking about how the dataset, architecture, loss function, and optimization method should be treated for the specific problem. Without such experience, I might still think of ML as either almighty black magic or obsolete, and probably all my ML-related astronomical works would be meaningless.

YUAN-SEN: I fear that the way that we are currently training students in astronomy might not be sufficient. To cultivate a well-versed community in both domains, we need a more flexible education system that will allow students to cross-pollinate between the two fields. This might mean we might want to consider other options in the PhD program, such as a joint PhD in astronomy and computer science that require students to publish not only in top astronomical journals but also in top machine learning conferences. I have seen such a movement at the Master level, which is certainly moving in the right direction.

LILIANNE: I believe that building the mathematical basis necessary for understanding what is behind ML implementations is more important than teaching ML code recipes. Teaching the foundations gives them freedom to learn (and produce!) more specific knowledge with quality, which could be ML or other things. I can give an example from my personal experience: my first contact with deep learning was through hands-on workshops. The analogies, the schema of neurons and layers, the frameworks, those were all well installed in my brain. However, I remember that I only truly understood deep learning years later after attending a 1-hour seminar where the speaker decided to show the mathematics behind it, with all the demonstrations. That relatively quick moment of realization was only possible because I spent a great amount of time attending fundamental courses. Looking at this question by other prism, I would actually give more emphasis on our failure in not cultivating stronger collaborations with machine learning researchers. Science is collaborative and we can make the best of it by working together, combining our expertise. One thing that can be harder than learning how to fit a ML model in Python is learning how to communicate with researchers from other STEM fields, as we all use different jargon.

2.7. *Ethics of AI in Astronomy*

SUNGWOOK: I agree that students need to learn the ethics of AI, mainly because students can develop their careers outside astronomy. However, I doubt if faculties in astronomy departments are best for teaching the ethics of AI. The ethics of AI rapidly changes over time mainly due to the various social issues such as human rights, which most astronomers do not (need to) care about much. Instead, since all fields will require AI as mandatory in the near future, I think a general AI ethics course should be open at the university level, not the department level.

LILIANNE: When we talk about Ethics of AI we are talking about the impact of AI and data-driven solutions on people's lives. From Coeckelbergh (2020),

> "AI ethics [...] is about the ethical challenges posed by current and near-future AI and its impact of our societies and vulnerable democracies. AI ethics is about the lives of people and it is about policy."

A classical example, based on the famous trolley problem, is the self-driving car's choice when faced with multiple, potentially lethal choices. Luckily for us astronomers, the worst we can do with AI in Astronomy is telling a galaxy "you are a star!". No celestial object's "lives" will ever be affected by our AI models. Actually, we have a bigger elephant in the room: everyone is susceptible to being harmed by the unethical behaviours of colleagues or supervisors. Unfortunately, the odds are higher when you are from a minority group. I sadly do not have enough fingers to count the number of astronomy students that I know that suffered from moral and/or sexual harassment in academia. Science is done by people, therefore ethics is involved in everything that we produce, not only where AI is involved. A better workplace in academia is something that we must strive for, thus ethics is something that we all must take lessons on.

YUAN-SEN: While deep learning has helped astronomy tremendously in the last few years, what perhaps has not been emphasised enough is that the astronomy dataset is the perfect dataset for the deep learning community to latch on to. On the one hand, underlining our data is the myriad of physical symmetries bound to challenge and advance our mathematical understanding of deep learning. On the other hand, in almost all cases, the astronomy dataset is free from many ethical concerns faced by the deep learning community. Making this point across will undoubtedly create a win-win situation for both communities.

References

Coeckelbergh, M. 2020, *AI Ethics*, The MIT Press, ISBN=9780262357067

Gatys, L. A., Ecker, A. S., & Bethge, M. 2016, *in 16th Annual Meeting of the Vision Sciences Society (VSS 2016)*, p. 326, Scholar One, Inc.

Gilpin, L. H., Bau, D., Yuan, B. Z., Bajwa, A., Specter, M. and Kagal, L. 2018, *The 5th IEEE International Conference on Data Science and Advanced Analytics (DSAA 2018)*

Hong, S. E., Jeong, D., Hwang, H. S., & Kim, J. 2021, *Astrophysical Journal*, 913, 76

Hong, S. E., Park, S., Jee, M. J., Bak, D., & Cha, S. 2021, *Astrophysical Journal*, 923, 266

Lin, T.-Y., Goyal, P., Girshick, R., He, K., & Dollár, P. 2017, *in IEEE International Conference on Computer Vision (ICCV)*, ed. K. Ikeuchi, G. Medioni, & M. Pelilo (New York: IEEE), 2999

Liu, G., Reda, F. A., Shih, K. J., Wang, T. C., Tao, A., & Catanzaro, B. 2018, *European Conference on Computer Vision*, pp. 89–105, Springer, Cham

Machine Learning in Astronomy: Possibilities and Pitfalls
Proceedings IAU Symposium No. 368, 2025
J. McIver, A. Mahabal & C. Fluke, eds.
doi:10.1017/S1743921323002892

Panel discussion: Methodology for fusion of large datasets

Nikhita Madhanpall[1], Kai Polsterer[2], Mike Walmsley[3] and Shay Zucker[4]

[1]IAU Office of Astronomy for Development, Cape Town, South Africa
email: nikhita@astro4dev.org

[2]Astroinformatics, HITS gGmbH, Heidelberg, Germany

[3]Jodrell Bank Center for Astrophysics, University of Manchester, Manchester, UK

[4]Department of Geophysics, Tel Aviv University, Tel Aviv, Israel

Abstract. During the IAU symposium, 368 "Machine Learning in Astronomy: Possibilities and Pitfalls" in Busan, we organized a panel discussion on the different aspects of data-fusion for large data-sets. Driven by the needs of the scientists, data-fusion technics had been introduces to enable multi-wavelength as well as multi-messenger approaches. This is necessary to get a more detailed and more complete representation of physical phenomena. We identified six different aspects related to data-fusion. Those aspects cover missing data, heterogeneous data, data-access in general, challenges related to data-size, FAIR-data, and future challenges.†

Keywords. methods: data analysis – methods: statistical

1. Introduction

Often, scientists strive for a more comprehensive description of physical phenomena, something essential to enable a deeper understanding of complex connections and correlations which a plain, spares, or simplified representation would not offer. To achieve this, multiple sources have to be combined; a process called data fusion. Especially when dealing with larger data sets, this process can play a crucial role to enable discoveries and allow for a better interpretation.

In astronomy, multi-frequency approaches were already early on, pushing the development of data fusion techniques. The formation of radio galaxies, cosmology studies in the direction of the Hubble tension, and linking gravitational waves to their electromagnetic counterparts are just examples of phenomena that benefit from data fusion approaches. Hereby, it is more than just linking data of different observatories together; it is furthermore the production of a more complete picture of our universe. With gravitational waves and astroparticle experiments providing different access to the universe, data-fusion became important to extended research in astronomy to multi-messenger aspects.

† This article is based on a panel discussion at the IAU Symposium 368 held in South Korea (2022) chaired by Kai Polsterer.

During the panel discussion, the following topics had been addressed:

Discussion

TOPIC 1 / MISSING DATA:

- How do you handle missing data?
- How is missing data replaced or filled in?
- What algorithms exist to deal with missing data?

TOPIC 2 / MERGING HETEROGENEOUS OBSERVATIONS:

- If data being merged come from different telescopes, different instruments, different filters, different noise/uncertainties, how to unify and merge such heterogeneous observations?

TOPIC 3 / DATA ACCESS:

- What are best practices to retrieve large datasets e.g. hundreds of thousands of image cutouts for DL tasks.
- How does one normalize them?
- How to handle pre-processing and post-processing?

TOPIC 4 / DATA SIZE:

- What to do if the data are simply too big?
- What are Methodologies for sampling, parallelization of algorithms, next generation research infrastructure, existing code/libraries for the data.

TOPIC 5 / FAIR DATA ACCESS:

- Organizing, structuring, and indexing/finding data in archives mainly by coordinates might be limiting.
- What might be other ways for a FAIR data access?
- Search by similarity, search by structure?
- Existing code/libraries for these?
- Science platforms

TOPIC 6 / FUTURE CHALLENGES:

- What are the upcoming challenges when going, e.g. from pre-processed catalogs to stream processing of data and events?
- How do the upcoming surveys handle this (SKA, LSST)?

Machine Learning in Astronomy: Possibilities and Pitfalls
Proceedings IAU Symposium No. 368, 2025
J. McIver, A. Mahabal & C. Fluke, eds.
doi:10.1017/S1743921323001229

The entropy of galaxy spectra

I. Ferreras[1,2], O. Lahav[2], R. S. Somerville[3] and J. Silk[4,5,6]

[1]Instituto de Astrofísica de Canarias, C/ Vía Láctea s/n, E38205, La Laguna, Tenerife, Spain
email: iferreras@iac.es

[2]Dept. of Physics and Astronomy, University College London, London WC1E 6BT, UK

[3]Center for Computational Astrophysics, Flatiron Institute, New York, NY 10010, USA

[4]Dept. of Physics and Astronomy, The Johns Hopkins University, Baltimore, MD 21218, USA

[5]Institut d'Astrophysique de Paris: CNRS&UPMC, Sorbonne University, F-75014, Paris, France

[6]Beecroft Institute, Dept. of Physics, University of Oxford, Keble Road, Oxford OX1 3RH, UK

Abstract. The ability of any Machine Learning method to classify the spectra of galaxies depending on the properties of the stellar component rests on the information content of the data. The well-known degeneracies found in population synthesis models suggest this information might be so entangled as to challenge the most sophisticated Deep Learning approaches. This contribution focuses on the traditional definition of entropy to explore this problem from a fundamental viewpoint. We find that the information content – when interpreting the spectrum as a probability distribution function – is reduced to a few spectral intervals that are strongly correlated. Dimensionality reduction via PCA suggests the standard 4000Å break strength and Balmer absorption are the two most informative regions in the analysis of galaxy spectra.

Keywords. methods: data analysis – methods: statistical – techniques: spectroscopic – galaxies: stellar content

1. Introduction

One of the potentially useful applications of machine learning (ML) to astrophysics concerns the analysis of the absorption features in galaxy spectra – we focus here on the rest-frame near ultraviolet/optical/near infrared. These features originate in the atmospheres of the stars that make up the stellar component, and encode information related to the evolution of galaxies. In broad terms, the spectral features can constrain the age and chemical composition of the underlying populations (see, e.g., Walcher et al. 2011), although the results suffer from degeneracies among the parameters that describe these populations, most notably age and metallicity (Worthey 1994). While the rest-frame optical spectrum of a galaxy consists of 100s to 1000s of flux measurements (over a range of wavelengths), the inherent constraining power is marginally better than photometric studies, that are reduced to 10s of flux measurements over the same spectral interval (Ferreras et al. 2022).

Machine learning methods can be applied to spectral data to extract information from the stellar populations, in a model-independent way, or at least can be trained on models to calibrate and optimise the derivation of model parameters. A number of papers have been already devoted to galaxy and stellar spectra (to name a few: Fabbro et al. 2018; Lovell et al. 2019; Ting et al. 2019; Wu & Boada 2019; Portillo et al. 2020; Liew-Cain et al. 2021), with overall rather generic results that, once more, do not give us a substantial advantage over high precision photometric studies. This study focuses on a fundamental question: *how much information can be extracted from spectra if we define them as probability distribution functions (PDFs)?*

A spectrum can be interpreted as a PDF if we think of the acquisition process as a photon counting experiment (Slonim et al. 2001), so that we can define the probability of detecting a photon from galaxy g with wavelength λ as $p(\lambda|g)$. Such an interpretation allows us to associate an entropy to the distribution, understood as a measure of surprise (Shannon & Weaver 1975):

$$H \equiv -\int p(\lambda)\log p(\lambda)d\lambda. \tag{1.1}$$

Probability distributions with high entropy will be highly undistinguishable from one another, whereas low entropy can, in principle, be interpreted as a high information content that could theoretically lead to the determination of detailed star formation and chemical enrichment histories from spectra. This result affects any ML method, as it inherently depends on the information content of the data. Thus, an entropy analysis will allow us to assess whether ML methods can ultimately solve one of the most complex inverse problems in astrophysics, namely, extracting the formation history of a galaxy from its spectrum.

Note that in the previous equation, the entropy is obtained as a single number for the whole spectrum. A more detailed quantity termed the entropy spectrum can be defined if we do the same analysis over a smaller (and fixed) spectral window ($\Delta\lambda$), centred at a location (λ_0) that varies over the available wavelength range. Moreover, we define the complement of this parameter, usually termed negentropy. Note that minimum information content corresponds to maximum entropy (or minimum negentropy):

$$\overline{\mathcal{H}}(\lambda_0) \equiv 1 + \int_{\lambda_0-\Delta\lambda/2}^{\lambda_0+\Delta\lambda/2} p(\lambda)\log p(\lambda)d\lambda. \tag{1.2}$$

In this definition, entropy is normalized so that the maximum possible value is set to 1. In galaxy spectra, the typical spectral resolution – mainly limited at the source by the velocity dispersion of the stellar component – is ~1-2Å, which motivates the typical resolution of spectrographs devoted to galaxies, such as the spectrograph that produced the 'legacy' Sloan Digital Sky Survey (SDSS) data (Smee et al. 2013). Previous studies show that an optimal window in this type of spectra is $\Delta\lambda$ ~100Å (Rogers et al. 2010; Hawkins et al. 2014).

2. Targeting optimal spectral windows from models

We use stellar population synthesis models (SPS) to identify the spectral regions where most of the information – from the point of view of entropy – is stored. These models are profusely applied to extract information from galaxy spectra and consist of three main ingredients (see, e.g., Conroy 2013) : 1) the distribution of stellar masses at birth, the so-called Initial Mass Function; 2) knowledge of stellar evolution that allows us to fully characterise a star of a given mass and chemical composition as it evolves with age; and 3) a stellar library that allows us to produce a synthetic spectrum as a superposition of stellar spectra, combined with knowledge of points 1 and 2 above. The basic unit in a population synthesis model is the simple stellar population (SSP) that corresponds to an ensemble of stars formed at the same time and with the same chemical composition. Complex mixtures, as those found in galaxies, are expected to correspond to superpositions of SSPs over a range of ages and chemical composition, weighted by some star formation and chemical enrichment history. Therefore, if we consider SPS models to give accurate representations of real galaxy spectra, we can use them to derive optimal

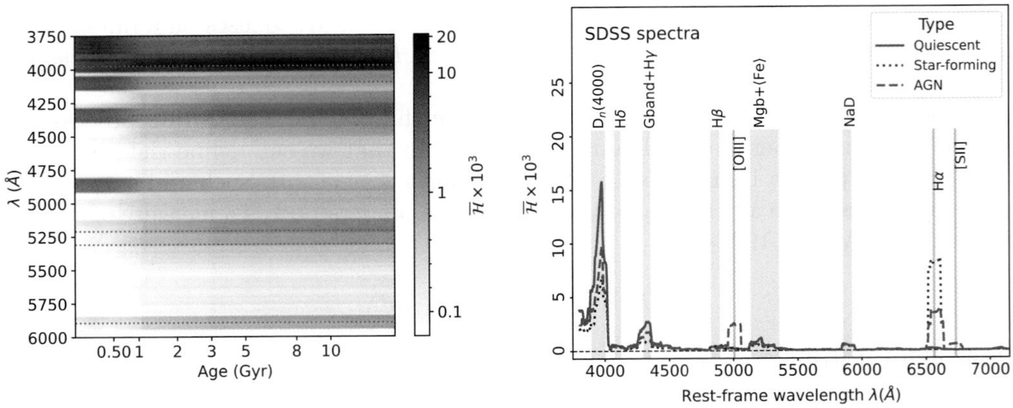

Figure 1. *Left:* 2D negentropy of a set of SSP models from Bruzual & Charlot (2003) at solar metallicity, for a range of stellar ages (horizontal axis), as a function of wavelength (left vertical axis). The greyscale bar on the right identifies the negentropy at a given position (age, wavelength). *Right:* Combined negentropy spectrum of a set of galaxy spectra from SDSS, split into star-forming (blue dotted), AGN (green dashed) and quiescent (red solid). Six spectral windows are defined to encapsulate most of the information content (shaded orange regions). The light blue vertical lines mark the positions of prominent emission lines.

spectral windows where the information content is highest. We adopt the definition in eq. 1.2 to find the negentropy spectrum, in intervals $\Delta\lambda = 100$Å. Fig. 1 (left) shows a 2D representation of the negentropy for the Bruzual & Charlot (2003) models with solar chemical composition, for a range of ages (horizontal axis), as a function of rest-frame wavelength (left vertical axis). The greyscale bar on the right maps the values of $\overline{\mathcal{H}}$. The most salient feature of this diagram is the inherently low information content, since $\overline{\mathcal{H}}$ is very small – noting that $\overline{\mathcal{H}} = 0$ means no information (maximum surprise) and 1 represents maximum information (no surprise, i.e. all photons having the same wavelength). Moreover, the number of regions with sizeable information content is restricted, and mostly concentrated on the bluer part of the spectrum. While harder to see, there is also a strong covariance among these different spectral windows.

3. The entropy of SDSS galaxy spectra

The analysis of the entropy spectra of synthetic data can be contrasted with real galaxy spectra. Fig. 1 (right) shows the negentropy of a set of over 76 thousand SDSS galaxy spectra (see Ferreras et al. 2022, for details). The spectra are classified according to their nebular emission properties into quiescent (red solid); star-forming (blue dotted); and AGN (green dashed). These three groups roughly map different key phases of galaxy evolution (see, e.g., Angthopo et al. 2019). Note that aside from the high negentropy of the emission lines – which originate in the diffuse gas in the galaxy and not its stellar component – the other sources of information are located in the regions suggested by the population synthesis models. The comparison motivates us to propose a small set of six spectral windows that encapsulate the majority of information, marked in Fig. 1 (right) as shaded boxes. It is not surprising that these windows eventually coincide with well-studied spectral features, especially the Lick index system (Trager et al. 1998).

4. The information of galaxy spectra

The six targeted regions are then explored with Principal Component Analysis (PCA), which produces linear combinations of these indices to create decorrelated data. PCA

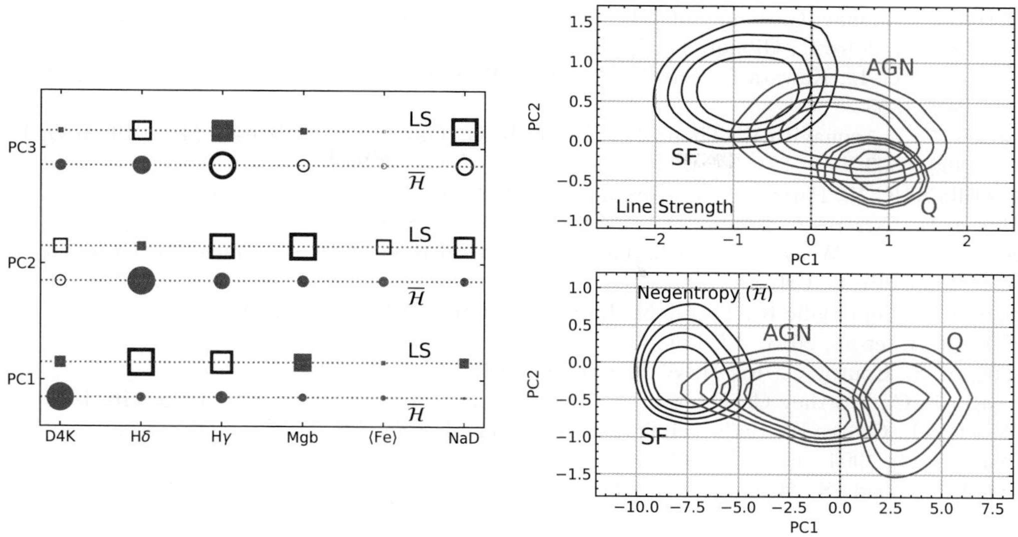

Figure 2. *Left:* Weights of the first three principal components (PC1, PC2, PC3), when applying PCA to SDSS galaxy spectra, using as input either line strengths (LS) or negentropy ($\overline{\mathcal{H}}$). The size of the symbols maps (linearly) the weight of each of the six spectral windows (horizontal axis). The red (solid) / blue (open) symbols correspond to a positive/negative sign of the weight. *Right:* Density plot with the projections of the SDSS spectra onto the first two PCs, split into star-forming (SF, blue), AGN (green) and quiescent (Q, red).

allows us to rank these combinations in decreasing amount of variance, which could be understood as information: those principal components (PCs) with low variance are expected to carry very little information concerning the properties of the stellar component. This is especially relevant for generic ML methods or full spectral fitting, where all the spectral data points are treated in the same way.

Fig. 2 (left) is a graphic representation of the weights of the six indices in the first three PCs. We compare two independent instances of PCA: *i)* using the actual line strengths (LS) as input, or *ii)* adopting the negentropy ($\overline{\mathcal{H}}$) as input to PCA. We note that the eigenvalues of the covariance (the 'scree plot'), concentrates a mugh higher fraction of total variance when using negentropy (94% of the total variance is represented by the first two PCs), with respect to line strengths (75% of total variance within PC1 and PC2). The first two PCs are dominated by the 4000Å break strength and Balmer absorption (especially Hδ). Other traditional spectral windows, such as Mgb and ⟨Fe⟩ have a substantially lower contribution, whereas Sodium absorption (NaD) contributes to the third PC. Fig. 2 (right) shows a density plot with the projection of the SDSS galaxy spectra onto the first two PCs, split into quiescent (Q, red), star-forming (SF, blue), and AGN (green), showing a clear transitioning status of the stellar content of AGN galaxies, between the star-forming stage and quiescence, and a clearer role of PC1 in this transitiom when PCA is applied to negentropy.

References

Angthopo J., Ferreras I., Silk J., 2019, *MNRAS Lett.*, 488, L99

Bruzual G., Charlot S., 2003, *MNRAS*, 344, 1000

Conroy C., 2013, *ARA&A*, 51, 393

Fabbro S., Venn K. A., O'Briain T., Bialek S., Kielty C. L., Jahandar F., Monty S., 2018, *MNRAS*, 475, 2978

Ferreras, I., Lahav, O., Somerville, R. S., Silk, J., 2022, RASTI, submitted, arXiv:2208.05489
Hawkins K., Jofré P., Gilmore G., Masseron T., 2014, *MNRAS*, 445, 2575
Liew-Cain C. L., Kawata D., Sánchez-Blázquez P., Ferreras I., Symeonidis M., 2021, *MNRAS*, 502, 1355
Lovell C. C., Acquaviva V., Thomas P. A., Iyer K. G., Gawiser E., Wilkins S. M., 2019, *MNRAS*, 490, 5503
Portillo S. K. N., Parejko J. K., Vergara J. R., Connolly A. J., 2020, *AJ*, 160, 45
Rogers B., Ferreras I., Peletier R., Silk J., 2010, *MNRAS*, 402, 447
Shannon C. E., Weaver W., 1975, The mathematical theory of communication, University of Illinois Press, Urbana
Slonim N., Somerville R., Tishby N., Lahav O., 2001, *MNRAS*, 323, 270
Smee S. A., et al., 2013, *AJ*, 146, 32
Ting Y.-S., Conroy C., Rix H.-W., Cargile P., 2019, *ApJ*, 879, 69
Trager S. C., Worthey G., Faber S. M., Burstein D., González J. J., 1998, *ApJS*, 116, 1
Walcher J., Groves B., Budavári T., Dale D., 2011, *Ap&SS*, 331, 1
Worthey G., 1994, *ApJS*, 95, 107
Wu J. F., Boada S., 2019, *MNRAS*, 484, 4683

Machine Learning in Astronomy: Possibilities and Pitfalls
Proceedings IAU Symposium No. 368, 2025
J. McIver, A. Mahabal & C. Fluke, eds.
doi:10.1017/S1743921323000728

Unsupervised classification: a necessary step for Deep Learning?

Didier Fraix-Burnet

Univ. Grenoble-Alpes / CNRS / IPAG, France
email: `didier.fraix-burnet@univ-grenoble-alpes.fr`

Abstract. Deep-learning algorithms have gained much popularity in the past decade. However, their supervised nature makes them fully dependent on the quality and completeness of the data samples used in the training processes. And when it comes to galaxy spectra for instance, there is simply no suited training sample available yet. This is where unsupervised classification (clustering) can come to the rescue.

Using the discriminant latent mixture-model based algorithm Fisher-EM, we 1) demonstrate the discriminative capacity of the method and its robustness with respect to noise on a sample of galaxy spectra simulated with the code CIGALE, 2) manage to classify ~700 000 spectra of galaxies from the SDSS, and 3) extend this classification to higher redshifts thanks to the VIPERS data (~80 000 spectra up to a redshift of 1.2). Finally, it appears that FisherEM is efficient for images of galaxies as well.

Keywords. Astrostatistics, Galaxies, Spectra, Unsupervised classification

1. Introduction

Classification of galaxies is often synonymous with the Hubble's morphological classification. The goal of modern extragalactic astronomy is to automatically classify billions of images in a very short time. Deep Learning sophisticated techniques, a priori perfectly suited for this task, rely on training samples that are based on this classification. Here comes a paradox: the current data are not only enormous in number but they are very much detailed, while the classes made by Hubble were based on a very limited information. In addition, the Hubble morphological classification is made by eye, a very lengthy and subjective procedure. The slightly more sophisticated classification by de Vaucouleurs (18 classes) is rarely used for the labelling of the training samples despite the fact that it corresponds better to modern detailed images. But this classification requires a very careful human examination with a high level of expertise, that Deep Learning is unable to reproduce.

We believe that our incredible data deserve better suited classification schemes, and the logical approach is to remain data-driven throughout the process. Noting that the Hubble classification was fully unsupervised, we promote the use of clustering (unsupervised classification) to describe the output of modern telescopes more faithfully and feed the supervised Artificial Intelligence techniques with more consistent labellings.

There is no global classification for the spectra of galaxies, apart from some taxonomy based on specific features. Somehow this is a chance to build something from scratch using unsupervised approaches. Our group is the first to propose an entirely data-driven atlas of galaxy spectra obtained from the whole optical spectra.

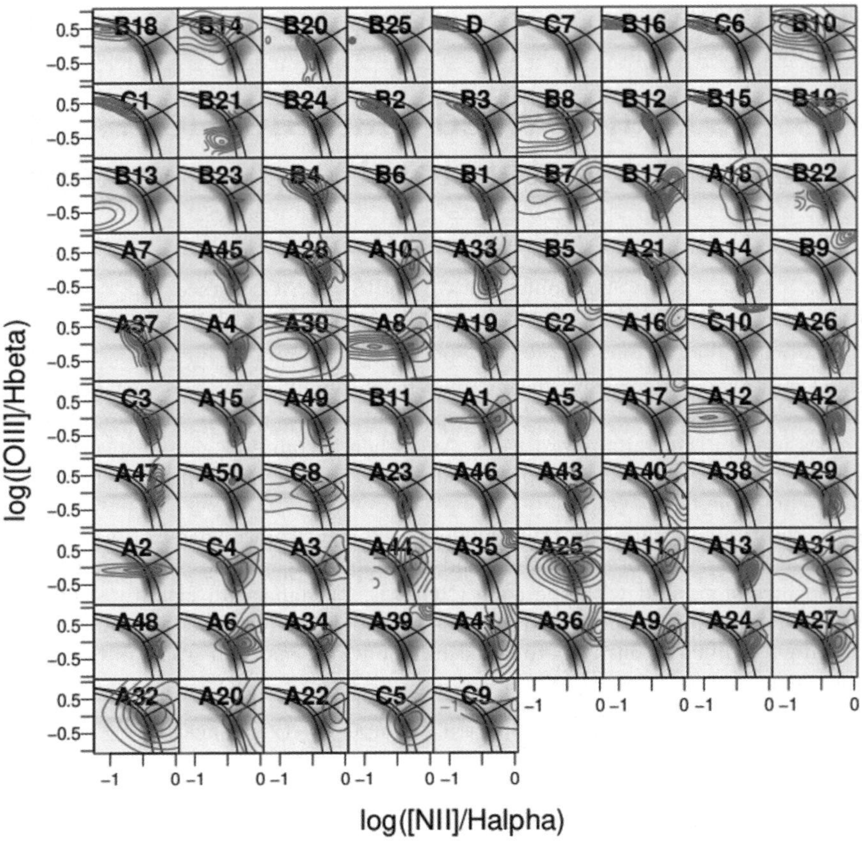

Figure 1. BPT diagram for the 86 classes of the SDSS samples of galaxy spectra ordered according to the slope of the mean spectra of each class.

2. Method: the algorithm FisherEM

A basic clustering technique is the k-means algorithm which tries to fit a mixture of identical Gaussians to a data set. Developments of this approach have introduced the possibility for the classes to have different distributions through the well-known optimisation procedure called expectation-maximisation (EM) iterations. Latent subspaces constitute further improvements for the clustering through a reduction of the dimensionality of the data set. The algorithm we use (FisherEM†; Bouveyron & Brunet 2012; Bouveyron & Brunet-Saumard 2014; available in R) introduces a supplementary step in the EM iteration to optimise the latent subspace based on the Fisher criterion. This Fisher criterion is the ratio between the variance between the classes to the variance within the classes. This guarantees that the latent subspace is optimised for the clustering goal.

3. Physical relevance of the classes

From a set of simulated spectra of galaxies obtained with the code CIGALE (Dubois et al. 2022), we have shown that each class corresponds to specific physical properties. Moreover, we show that the classification is robust with respect to the level of signal to noise ratio, up to three. But even at a signal to noise ratio of one, the obtained

† `https://www.rdocumentation.org/packages/FisherEM/versions/1.6`

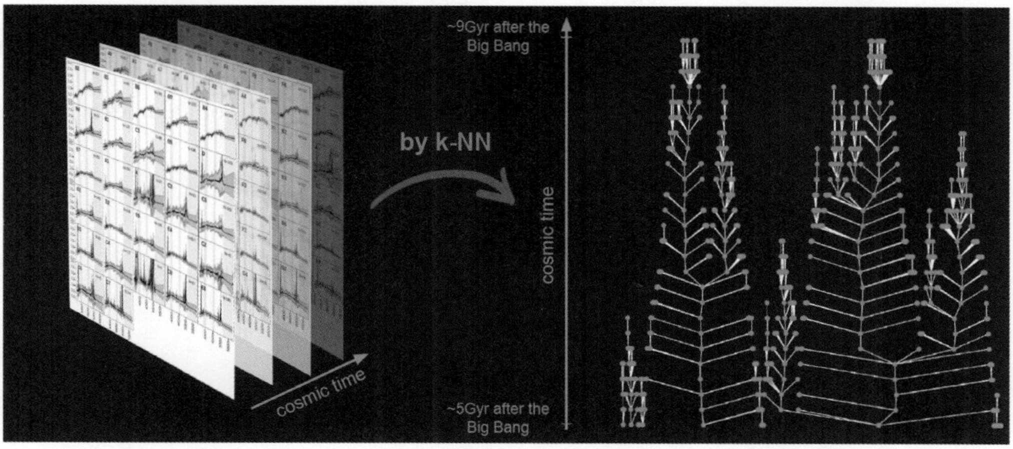

Figure 2. Tree of classes of galaxy spectra for the VIPERS sample. Courtesy Julien Dubois.

classification is consistent despite being less detailed. This result gives a lot of confidence in the capabilities of the algorithm FisherEM.

4. An atlas of SDSS galaxy spectra

We performed an unsupervised classification of 702248 spectra of galaxies and quasars with redshifts smaller than 0.25 that were retrieved from the Sloan Digital Sky Survey (SDSS) database, release 7 (Fraix-Burnet et al. 2021). The optimum number of classes given by a penalised likelihood criterion called Integrated Complete Likelihood (ICL) is 86 classes, of which the 37 most populated ones gather 99% of the sample.

A preliminary physical interpretation of the classes (Fraix-Burnet et al. 2021) show that the spectra are well discriminated by their slope, but more remarkably by their emission line ratios as illustrated by the BPT diagram of $[OIII]/H_\beta$ versus $[NII]/H_\alpha$ (Fig. 1).

This is the first time that an automatic, objective and robust unsupervised classification is established on such a large number of galaxy spectra. We are currently working on the detailed interpretation of each of the classes in order to provide a physical atlas of galaxy spectra at low redshift.

5. Extending the classification up to redshift 1.2

Going to higher redshifts is difficult because the information contained in the spectra varies. The only solution is to divide the samples into bins of redshift, compromising between the number of spectra and the number of common data points. We have succeeded in obtaining an unsupervised classification for each of the 26 bins between redshifts 0.4 to 1.2 on a sample of 80 000 spectra from the VIPERS survey†. We then used the k-Nearest Neighbour (kNN) technique to connect the classes between two consecutive bins to provide an "evolutionary" tree (Fig. 2, Dubois et al., in prep).

6. Conclusion

Our results can be used in several ways. Firstly, we provide a data-driven atlas of galaxy spectra with its evolution in redshift. Secondly, the statistical mixture-based model can be used to instantaneously classify any new spectra by simply computing the probabilities

† `http://www.vipers.inaf.it/`

for this spectrum among all the classes. Lastly, our classification makes a data-driven training sample of nearly 800 000 optical spectra of galaxies up to a redshift of 1.2 that can be used for instance in Deep Learning approaches.

We finally note that we have obtained excellent results on images of galaxies with this same algorithm FisherEM.

References

Bouveyron, C., Brunet, C. 2012, *Statistics and Computing*, 22, 301
Bouveyron, C., Brunet-Saumard, C. 2014, *Computational Statistics*, 29, 489
Dubois, J., Fraix-Burnet, D., Moultaka, J., *et al.* 2022, *A&A*, 663, A21
Fraix-Burnet, D., Bouveyron, C., Moultaka, J. 2021, *A&A*, 649, A53

Machine Learning in Astronomy: Possibilities and Pitfalls
Proceedings IAU Symposium No. 368, 2025
J. McIver, A. Mahabal & C. Fluke, eds.
doi:10.1017/S1743921323001308

Spectral identification and classification of dusty stellar sources using spectroscopic and multiwavelength observations through machine learning

Sepideh Ghaziasgar[1], **Amirhossein Masoudnezhad**[2], **Atefeh Javadi**[1], **Jacco Th. van Loon**[3], **Habib G. Khosroshahi**[1] and **Negin Khosravaninezhad**[2,4]

[1]School of Astronomy, Institute for Research in Fundamental Sciences (IPM), P.O. Box 19395-5531, Tehran, Iran
email: sepide.ghaziasgar@gmail.com

[2]Department of Physics, Sharif University of Technology, P.O. Box 11155-9161, Tehran, Iran

[3]Astrophysics Group, Lennard-Jones Laboratories, Keele University, ST5 5BG, UK

[4]Department of Physics and Astronomy, University of California Riverside, CA 92521, USA

Abstract. We proposed a machine learning approach to identify and distinguish dusty stellar sources employing supervised methods and categorizing point sources, mainly evolved stars, using photometric and spectroscopic data collected over the IR sky. Spectroscopic data is typically used to identify specific infrared sources. However, our goal is to determine how well these sources can be identified using multiwavelength data. Consequently, we developed a robust training set of spectra of confirmed sources from the Large and Small Magellanic Clouds derived from SAGE-Spec Spitzer Legacy and SMC-Spec Spitzer Infrared Spectrograph (IRS) spectral catalogs. Subsequently, we applied various learning classifiers to distinguish stellar subcategories comprising young stellar objects (YSOs), C-rich asymptotic giant branch (CAGB), O-rich AGB stars (OAGB), Red supergiant (RSG), and post-AGB stars. We have classified around 700 counts of these sources. It should be highlighted that despite utilizing the limited spectroscopic data we trained, the accuracy and models' learning curve provided outstanding results for some of the models. Therefore, the Support Vector Classifier (SVC) is the most accurate classifier for this limited dataset.

Keywords. stars: classification, stars: AGB and post-AGB, stars: dust, galaxies: spectroscopic catalog, galaxies: multiwavelength observations, methods: data analysis

1. Introduction

Surveys have acquired large datasets over the multiwavelength sky. Understanding dusty stellar sources, such as young stellar objects (YSOs) and evolved stars, has been dramatically enhanced by spectroscopic data acquisition and analysis. Indeed, the growing number of spectroscopic observations and confirmed stellar spectra has necessitated the development of novel approaches to identifying and classifying stellar sources. As a result, machine learning enhances the accuracy of spectral classification of dusty point sources while also validating candidate spectroscopic catalogs and identifying unclassified spectra.

Table 1. Number of samples.

Classes	LMC	SMC	Total
YSO	213	51	264
C-AGB	152	39	191
O-AGB	89	19	108
RSG	72	22	94
Post-AGB	33	4	37

For probing the Magellanic Clouds, photometric and spectroscopic classification of star properties and content is used. Although photometric approaches are faster, spectroscopic methods provide highly accurate analysis of large samples. Indeed, the Magellanic Clouds are suited to investigating stellar contributions to dust emission (Ruffle et al. 2015). An early-evolutionary star is known as a YSO. Due to star formation, YSOs are surrounded by gas and dust disks (Suh 2016). Therefore, YSOs contribute to our understanding of star formation and allow us to characterize star-forming regions. Mass, age, and the star-forming environment determine YSO properties. In addition, evolved stars are dust producers, and their role in enriching galaxies is crucial.

In the final stage of stellar evolution, low- and intermediate mass (0.8-8 $M_\odot$) stars evolve into the Asymptotic Giant Branch (AGB) phase, and high mass ($M \geq 8$ $M_\odot$) stars enter the Red Supergiants (RSG) phase. Most AGB stars are long-period variables (LPVs) with large amplitude pulsations; and they have circumstellar dust envelopes with high mass-loss rates (Saremi et al. 2020; Navabi et al. 2021; Javadi & van Loon 2022). Dust production and mass loss rates of pulsating AGB stars and RSG stars are determined by evolved stars. AGB stars are classified as O-rich AGB (OAGB) or C-rich AGB (C-AGB) based on the chemistry of the photosphere and the outer dust envelope (Suh 2020, 2021; Javadi et al. 2011a, 2011b, 2013). We can detect these sources in the infrared band due to AGB stars and RSGs are luminous ($\sim 10^{3.5-5.5}$ $L_\odot$) and red (Javadi & van Loon 2022). Post-AGB stars are luminous, low- or intermediate-mass stars, which end their AGB branch evolution by a phase of mass loss (Van Winckel 2003). Consequently, infrared (IR) spectral observations and classification are essential for studying YSOs and evolved stars (Boyer et al. 2017).

2. Data

We used tabular FITS datasets available from the Strasbourg astronomical Data Center (CDS) for the Large and Small Magellanic Clouds taken from SAGE-Spec Spitzer Legacy and SMC-Spec Spitzer Infrared Spectrograph (IRS) spectral catalogs (Jones et al. 2017a; Ruffle et al. 2015). SAGE is the project for Surveying the Agents of Galaxy Evolution for tracking dust and gas in Magellanic Clouds (Meixner et al. 2006; Woods et al. 2011; Ruffle et al. 2015; Jones et al. 2017a). To prepare the final data for implementation in various models, we performed several steps of preprocessing, including label encoding and data imputation. Following that, we selected 22 features including magnitude and flux at different wavelengths, and 694 samples were collected from both catalogs. Consequently, we allocated a feature data frame (X) with 694 rows and 22 columns and a target array (Y) for each sample's related class. We used some features comprising Umag, Bmag, Vmag, Imag, <Vmag>, <Imag>, J2mag, H2mag, Ks2mag, Jmag, Hmag, Ksmag, 3.6 μm, 4.5 μm, 5.8 μm, 8.0 μm. Table 1 provides a list of stellar classes.

3. Method

Machine Learning (ML) deals with techniques and methods that enable machines to learn. In this regard, algorithms are developed to perform tasks and construct models to

Table 2. Performance of Classifiers: This table compares the performance of different classification models based on f1-score (%) and accuracy (%). The Support Vector Classifier (SVC) is the most accurate classifier for this limited dataset having 90 percent accuracy and an average f1-score of 88 percent.

Classifier	Average f1-score (%)	Accuracy (%)
K-Nearest Neighbors	79	86
Decision Tree Classifier	80	86
Linear Discriminant Analysis	86	89
Gaussian Naive Bayes	77	81
Stochastic Gradient Descent	87	90
Support Vector Classification	88	90

make inferences depending on the data type. Moreover, these inferences can be a pattern, description, or prediction based on past data (Sturrock et al. 2019). Training, validation, and testing datasets are commonly used in model development. Machine learning methods are applied to astronomy thanks to big astronomical data from ground-based and space telescopes. Classification is one of the areas where astronomers are interested in implementing supervised and unsupervised algorithms (Ivezić et al. 2014). To categorize dusty stellar sources, we used K-Nearest Neighbor (KNN), Decision Tree Classifiers (DTC), Linear Discriminant Analysis (LDA), Gaussian Naive Bayes, Stochastic Gradient Descent (SGD), and Support Vector Classification (SVC) for limited spectral data.

To understand some results in the following, some standard definitions and functions should be briefly explained.

Classification report; contains the model's precision, recall, and f1-score values for each class. Precision; is the fraction of relevant instances among the retrieved ones. Recall; is the fraction of relevant cases retrieved, and in the following formula, f1-score is a combination of precision and recall.

$$f_1\text{-}score = \frac{2 \times precision \times recall}{precision + recall}$$

Accuracy; As a metric for evaluating classifiers, the number of correct predictions divided by the total number of predictions.

Validation Curve; This curve determines and plots training and test scores for different values of classifier hyperparameters.

Learning Curve Function; This function splits the dataset into training and test data. It determines the cross-validated training and test scores to demonstrate how the classifier's performance changes with varying training samples. The learning curve optimizes algorithms and their parameters over time as they are learned gradually.

Confusion Matrix; This is an N x N table layout that visualizes an algorithm's performance in which rows represent instances in the actual class, and the columns represent the predicted class.

4. Some Results and On-going work

We implemented several models, as shown in Table 2. As an example, in the following paragraph, we presented the Support Vector Classifier (SVC) as a supervised learning method for classification, which works by mapping data to a high-dimensional feature space so that data points can be classified even when data are not otherwise linearly separable (Kovács, Szapudi 2015; Krakowski et al. 2016).

In this study, we tuned the values of SVC hyperparameters (e.g., C, gamma), and based on validation curves for hyperparameters, C should be more than 100 and gamma

Table 3. In the classification report, the f1-score is calculated by averaging. The macro average f1-score represents an average of f1-score over classes. The weighted average f1-score is calculated as the mean of all per-class f1-scores while considering each class's support. Based on true occurrences of a category, support is determined.

	precision	recall	f1-score	support
CAGB	0.95	1.00	0.97	37
OAGB	0.80	0.77	0.78	26
PAGB	0.88	0.88	0.88	8
RSG	0.88	0.75	0.81	20
YSO	0.92	0.96	0.94	48
accuracy			0.90	139
macro avg	0.89	0.87	0.88	139
weighted avg	0.90	0.90	0.90	139

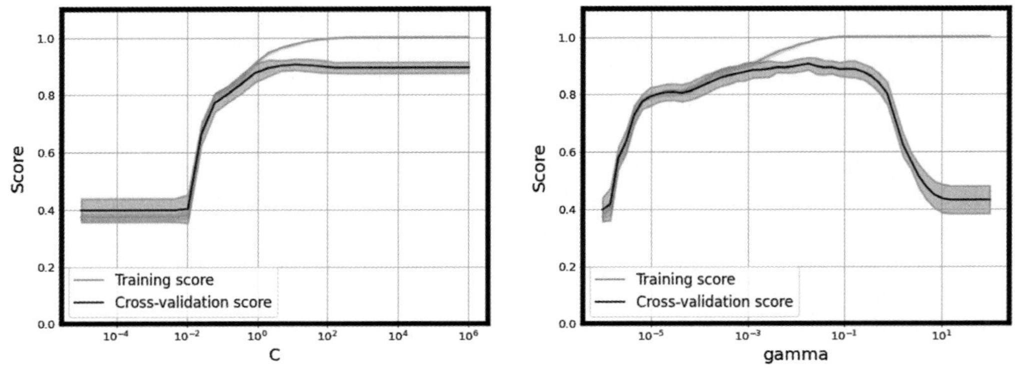

Figure 1. Validation Curve of the Support Vector Classifier for C and gamma hyperparameters. These SVC hyperparameter values are tuned based on validation curves. As shown, C is more than 100 and gamma ~ 0.1.

~ 0.1, respectively (Figure 1). The final results showed 90 percent accuracy and an average of 88 percent for the f1-score, which were the most accurate among all models. Therefore, SVC is the most accurate classifier for our dataset without augmentation, as shown in Table 3. The confusion matrix demonstrates the SVC model's performance in stellar classifications (Figure 2). A Support Vector Classifier also learns gradually through proper training samples, as demonstrated in Figure 3.

In the future, since the lack of data and its imbalance resulted in lower classification scores in some classes (e.g., PostAGBs), we will apply augmentation to our dataset to improve our accuracy. We are looking for an optimal model for our data augmentation and implementing it to classifiers to demonstrate the models' performance before and after augmentation. One way to solve this problem is to oversample the minority class data, which can be achieved by synthesizing samples from the minority classes before fitting a model. To further investigate, we aim to complete this study for machine learning models to categorize dusty stars that can improve by employing data augmentation and expanding spectroscopic data and approved catalogs, including the JWST and upcoming spectroscopic survey data. A mid-infrared stellar population study of galaxies is generally limited to nearby galaxies, whereas JWST will allow studies of galaxies within the Local Volume (Jones et al. 2017b). Additionally, we will be able to enhance the functionality and accuracy of our ML models by training sophisticated classifiers.

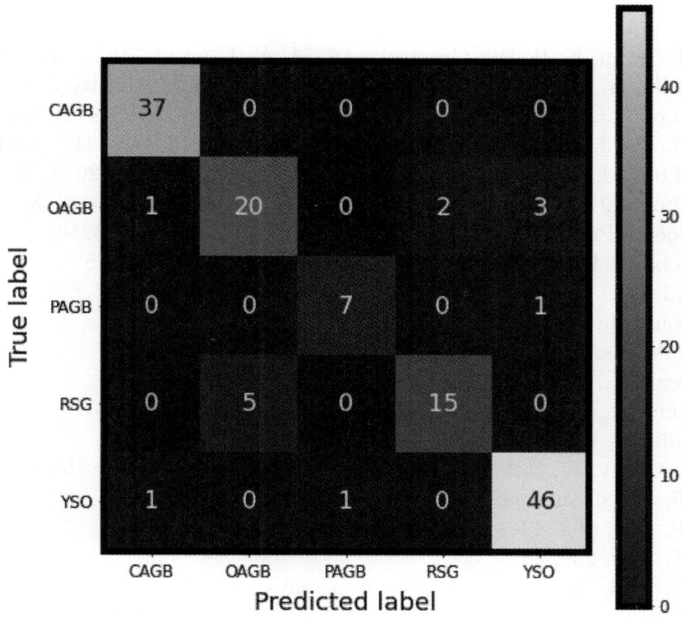

Figure 2. Confusion Matrix of the Support Vector Classifier: This layout shows how well a classification algorithm performs. Based on the model, it predicts that a certain number of objects belong to each class. The dusty stellar class diagonals show the performance of predicted and true labels.

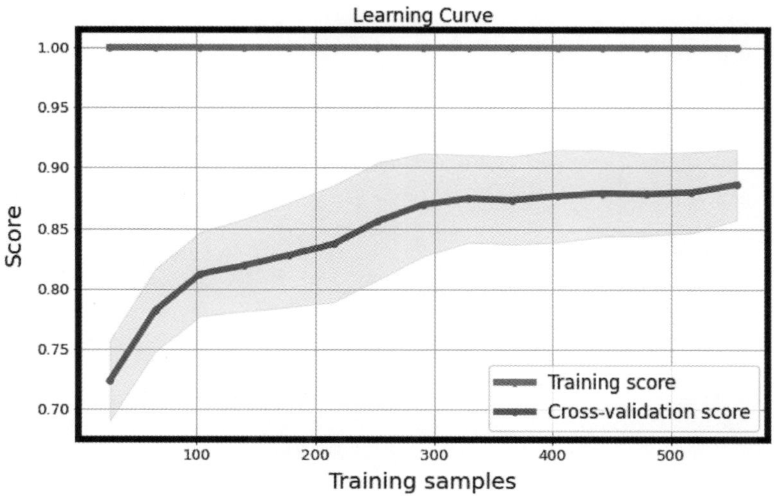

Figure 3. Learning Curve of the Support Vector Classifier: This curve shows that learning increases gradually and optimal training samples fit the model. The red (straight line) represents the Training score, while the green (curved line) indicates the Cross-validation score.

Acknowledgements

The authors thank the School of Astronomy at the Institute for Research in Fundamental Sciences (IPM) and the Iranian National Observatory (INO) for supporting this research. S. Ghaziasgar is grateful for the IAU support.

References

Boyer, M. L., McQuinn, K. B. W., Groenewegen, M. A. T., et al. 2017, ApJ, 185, 152
Ivezić, Ž., Connolly, A., Vanderplas, J., et al. 2014, Princeton University Press
Javadi, A., van Loon, J.Th., Mirtorabi, M.T., et al. 2011a, MNRAS, 411, 263
Javadi, A., van Loon, J.Th., Mirtorabi, M.T., et al. 2011b, MNRAS, 414, 3394
Javadi, A., van Loon, J.Th., Khosroshahi, H., Mirtorabi, M.T., et al. 2013, MNRAS, 432, 2824
Javadi, A., van Loon, J.Th., 2022, Proceedings of IAU Symposium, 366, arXiv:2204.08944
Jones, O. C., Woods, P. M., Kemper, F., et al. 2017a, MNRAS, 470, 3250
Jones, O. C., Meixner, P. M., Justtanont, k., et al. 2017b, ApJ, 841, 15
Krakowski, T., Mal ek, K., Bilicki, M., et al. 2016, A&A, 596, A39
Kovács, A., Szapudi, I., Bilicki, M., et al. 2015, MNRAS, 448, 1305
Meixner, O. C., Woods, P. M., Kemper, F., et al. 2006, AJ, 132, 2268
Navabi, M., Saremi, E., Javadi, A., et al. 2021, ApJ, 910, 127
Ruffle, M. E., Kemper, F., Jones, O. C., et al. 2015, MNRAS, 451, 3504
Saremi, E., Javadi, A., van Loon, et al. 2020, ApJ, 894, 135
Sturrock, G. C., Manry, B., Rafiqi, S., et al. 2019, SMU Data Science Review, 2, 1
Suh, K.W., 2016, J. Astron. Space Sci., 33, 119
Suh, K.W., 2020, ApJ, 891, 43
Suh, K.W., 2021, ApJ, 256, 43
Van Winckel, H., 2003, Annu. Rev. Astron. Astrophys., 41, 391
Woods, P. M., Oliveira, J. M., Kemper, F., et al. 2011, MNRAS, 411, 1597

Machine Learning in Astronomy: Possibilities and Pitfalls
Proceedings IAU Symposium No. 368, 2025
J. McIver, A. Mahabal & C. Fluke, eds.
doi:10.1017/S1743921323001047

Simulating transient burst noise with gengli

Melissa Lopez[1,2], Vincent Boudart[3], Stefano Schmidt[1,2] and Sarah Caudill[1,4,5]

[1]Institute for Gravitational and Subatomic Physics (GRASP), Utrecht University, Princetonplein 1, 3584 CC Utrecht, The Netherlands
email: `m.lopez@uu.nl`

[2]Nikhef, Science Park 105, 1098 XG, Amsterdam, The Netherlands

[3]STAR Institute, Bâtiment B5, Université de Liège, Sart Tilman B4000 Liège, Belgium

[4]Department of Physics, University of Massachusetts, Dartmouth, MA 02747, USA

[5]Center for Scientific Computing and Visualization Research, University of Massachusetts, Dartmouth, MA 02747, USA

Abstract. In the field of gravitational-wave (GW) interferometers, the most severe limitation to the detection of transient signals from astrophysical sources comes from transient noise artefacts, known as glitches, that happens at a rate around 1 per minute. Because glitches reduce the amount of scientific data available, there is a need for better modelling and inclusion of glitches in large-scale studies, such as stress testing the search pipelines and increasing the confidence of detection. In this work, we employ a Generative Adversarial Network (GAN) to produce a particular class of glitches (*blip*) in the time domain. We share the trained network through a user-friendly open-source software package called `gengli` and provide practical examples of its usage.

Keywords. Generative adversarial networks, gravitational waves, synthetic data, machine learning.

1. Introduction

The existence of gravitational-wave (GW) signals was successfully proven by LIGO and Virgo collaborations during the first observing run (O1) (B. P. Abbott *et al.* 2016). After an upgrade of the detectors to increase their sensitivity, Advanced LIGO (J. Aasi *et al.* 2015) started in November 2016 the second observing run (O2), which Advanced Virgo (F. Acernese *et al.* 2015) joined in August 2017 (B. P. Abbott *et al.* 2017). Following significant upgrades, in April 2019, the third observing run (O3) was initiated by LIGO-Virgo collaboration (B. P. Abbott *et al.* 2020; R. Abbott *et al.* 2021). In the coming years, the improved second generation of interferometers and the construction of the third generation of detectors, such as Einstein Telescope (S. Hild *et al.* 2011; M. Maggiore *et al.* 2020), will increase significantly the detection sensitivity.

Despite the significant improvements to isolate the detectors from non-cosmic disturbances, they are still susceptible to non-Gaussian noise, known as "glitches", which come in a wide variety of time-frequency morphologies and are produced by instrumental or environmental causes. They reduce the amount of analyzable data, biasing astrophysical detection, and mimicking GW signals (B. P. Abbott *et al.* 2016), as in Fig. 1. Thus, it is fundamental to identify and characterize them for their elimination.

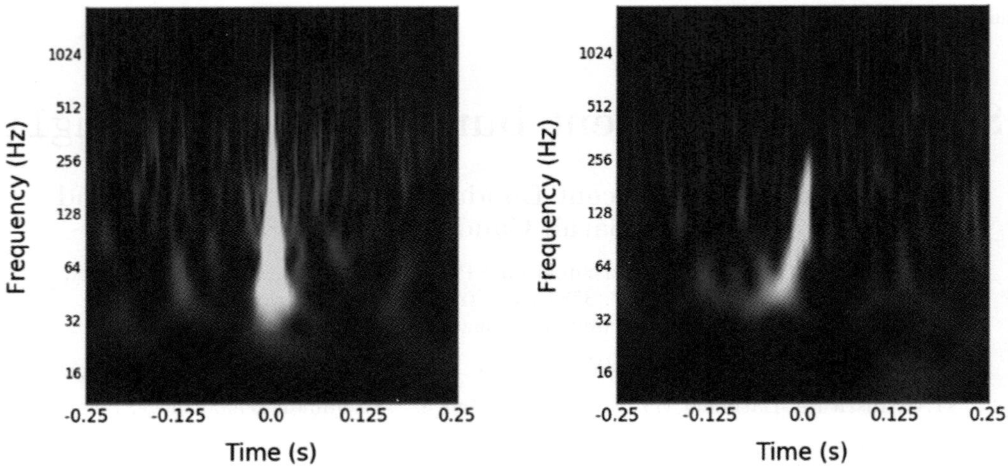

Figure 1. A blip glitch *(left)* that is similar to an event with total mass $106.6^{+13.5}_{-14.8} M_{\odot}$ *(right)*.

Due to the overwhelming amount of glitches present in the LIGO data (R. Abbott *et al.* (2021)), identifying them goes beyond human ability. An exciting solution is to construct machine learning (ML) algorithms to classify their different morphologies. With this idea in mind, M. Zevin *et al.* (2017) combine the strengths of both humans and computers to analyze and characterize LIGO glitches. Through the Zooniverse platform, volunteers provide large labelled data sets to train an ML algorithm, called *Gravity Spy*, while the ML algorithm learns to classify the rest of the glitches correctly and provides feedback to the participants. In practice, we feed to the ML classifier the glitch time series that we wish to classify. The Q-transform of the input is created and fed to the ML classifier, that assigns a label and a confidence value c_{GS}, where c_{GS} is the classification probability (see M. Zevin *et al.* (2017)).

While the identification of glitches is the first step towards their robust mitigation, in this investigation we intend to *generate* the known classes of glitches with Generative Adversarial Networks (GAN) to further understand their principal features. This does not only allow us to enhance our understanding of their morphologies but it can also be used for various applications in GW data analysis, such as mock data challenges (see M. Lopez *et al.* (2022) for a discussion). In this work, that accompanies the main publication M. Lopez *et al.* (2022), we present a flexible and user-friendly tool *gengli*† for glitch generation.

2. Data

To train our ML model, we select blips from Livingston (L1) and Hanford (H1) that have a confidence $c^1_{GS} \geq 0.9$, during O2‡. However, since the glitch is surrounded by non-stationary and uncorrelated noise, there is little structure that our ML method can retrieve. Therefore, we need to extract glitches from the input strain. For this aim, we employ BayesWave (N. J. Cornish and T. B. Littenberg 2015) to fit and reconstruct the input signal with data-driven wavelet models. The *whitening* operation is defined on the frequency domain glitch $\tilde{g}$ as: $\tilde{g}_W = \tilde{g}/S_n(f)$, where $S_n(f)$ is the power spectral density (PSD) of the input data.

The input provided to BayesWave is a time series of 2.0 s at 4096 Hz, where the blip is centred. After the reconstruction, we crop the output to a size of 938 data points,

† The code is released as a Git repository in melissa.lopez/gengli
‡ The data is accessible through the GWOSC (`https://www.gw-openscience.org/data/`)

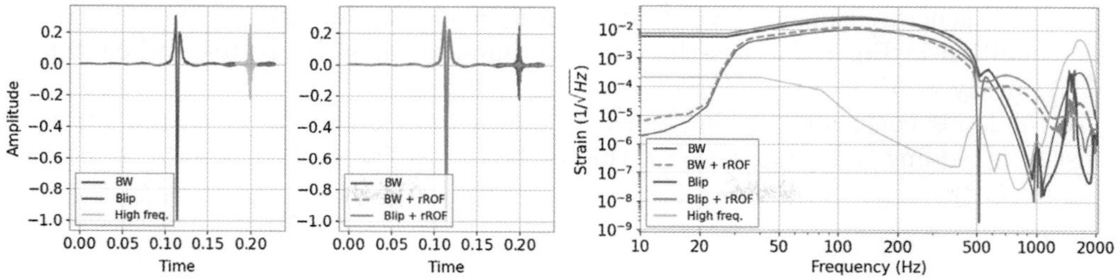

Figure 2. *(Left)* Reconstructed blip with BayesWave. *(Middle)* Denoised reconstructed blip with rROF. *(Right)* Resulting amplitude spectral density (ASD) for reconstructed and denoised blips.

removing irrelevant data. We evaluate the quality of the reconstruction by injecting the output in real whitened noise, and re-evaluating with Gravity Spy to select blip glitches with $c^2_{GS} \geq 0.9$.

The reconstruction of BayesWaves is not perfect since there is still some noise contribution at high frequencies (light blue) that will hinder the learning of our ML algorithm, (see Fig. 2 *(left)*). To minimize this contribution we employ regularized Rudin-Osher-Fatemi (rROF) proposed in A. Torres *et al.* (2014), which solves the denoising problem as a variational method. In Fig. 2 (middle), we plot the BayesWaves reconstruction denoised with rROF (dashed orange), and the denoised characteristic blip (green). In Fig. 2 (right), we show the amplitude spectral density (ASD) of the BayesWaves reconstruction with and without denoising (grey and dashed orange), as well as the characteristic peak with and without denoising (blue and green) and the original high-frequency contribution (light blue). We can maintain the structure of the characteristic peak by damping the power of the high-frequency contribution.

3. Methodology

3.1. *Generative Adversarial Networks*

GANs, first introduced by I. J. Goodfellow *et al.* (2014), are a class of generative algorithms in which two networks compete against each other to achieve realistic data generation. One network is responsible for the generation of new data from random noise, while the other tries to discriminate the generated samples from the real training data. The generator learns the features of real data that should be mimicked to "foul" the discriminator, and stores them in a latent space. At the end of the training, new samples are drawn by randomly taking a latent space vector and passing it to the generator, which translates this into real data.

The work from I. J. Goodfellow *et al.* (2014) suffers from two major problems, known as *vanishing gradients* and *meaningless loss function* (L. Weng 2019). To address those issues M. Arjovsky *et al.* (2017) developed Wasserstein GANs, where they use the Earth's mover distance or Wasserstein-1 distance (W_1) as a loss function. This change led to reformulate the optimization problem as $\theta_{opt} = arg\,\underset{\theta}{min}\, W_1(P_x \| P_{\tilde{x}})$, where W_1 is evaluated between the true data distribution P_x and the fake data distribution $P_{\tilde{x}}$. Rewriting this equation yields,

$$\theta_{opt} = arg\,\underset{\theta}{min}\,\underset{\phi:||D(x,\phi)||_L \leq 1}{max} L(\phi,\theta) \quad \text{with } L(\phi,\theta) = -E_{x\sim P_x}\big[D(x,\phi)\big] + E_{\tilde{x}\sim P_{\tilde{x}}}\big[D(\tilde{x},\phi)\big] \tag{1}$$

where $D(\phi)$ and $G(\theta)$ are for the discriminator and the generator is seen as a function of their weight. $E_{x\sim P_x}$ indicates that the expression has been averaged over a batch of real images. Similarly, $E_{\tilde{x}\sim P_{\tilde{x}}}$ indicates that the expression has been average over a batch of generated images, where $\tilde{x} = G(z, \theta)$ and z is the latent space vector. The new condition in Eq. 1 imposes that the discriminator must be 1-Lipschitz continuous (M. Arjovsky *et al.* 2017).

To fulfil this constraint, we implement the idea by I. Gulrajani *et al.* (2017), which consists in adding a regularization term to the discriminator loss, known as *gradient penalty* (GP):

$$L_{tot} = L(\phi, \theta) + \lambda\, GP(\phi) \; with \; GP(\phi) = E_{\hat{x} \sim P_{\tilde{x}}} \left[\left(|| \nabla_x \; D(\hat{x}, \phi) ||_2 - 1 \right)^2 \right], \tag{2}$$

where λ is a regularization parameter, $|| \cdot ||_2$ denotes the L2-norm and $\hat{x}$ is evaluated following: $\hat{x} = \tilde{x}\, t + x\, (1 - t)$, being x a real sample and $\tilde{x}$ a fake sample, with t sampled uniformly $\in [0, 1]$. However, GP cannot penalize W_1 everywhere. In particular, at the beginning of the training, being the generated samples quite far from the true data manifold, the Lipschitz condition is not enforced until the generator becomes sufficiently good.

To overcome this obstacle, X. Wei *et al.* (2018) have proposed a second penalization term that directly penalises the points near any observed real data point x. Therefore, when the gradient penalty fails to enforce the Lipschitz continuity in the close vicinity of x, the new term will constrain the latter. To penalize the real data manifold, X. Wei *et al.* (2018) applied their new constraint to two perturbed versions of the real samples x. For this, they introduced dropout layers into the discriminator architecture. The features kept by the dropout layer are random, which ultimately leads to two different estimates noted $D(x')$ and $D(x'')$. The penalty, called *consistency term* (CT), is then applied to these two estimates following:

$$CT(\phi) = E_{x \sim P_x} \left[max\left(0, \; d(D(x', \phi), D(x'', \phi)) + 0.1\, d(D\_(x', \phi), D\_(x'', \phi)) - M' \right) \right] \tag{3}$$

where d(.,.) is the L2 metric, $D\_$ stands for the second-to-last layer output of the discriminator and M' is a constant value. The final discriminator loss in X. Wei *et al.* (2018) is then:

$$L_D = L(\phi, \theta) + \lambda_1\, GP(\phi) + \lambda_2\, CT(\phi) \tag{4}$$

with λ_2 being the consistency parameter, that is used to tune the weight of CT for GP. While in Eq. 4 we update the weights of the discriminator, we update the weights of the generator as in M. Arjovsky *et al.* (2017), with the following expression:

$$L_G = -E_{\tilde{x} \sim P_{\tilde{x}}} \left[D(\tilde{x}, \phi) \right] \tag{5}$$

3.2. *Architecture and training procedure*

To build our model we use convolutional neural networks (CNN) in one dimension. In the generator, we make use of nearest-neighbour upsampling layers to avoid artefacts. We set the kernel size $k = 5$, padding $p = 0$, stride $s = 1$, with increasing dilation to enlarge the receptive field. Batch normalization is applied after each layer except for the penultimate one. Finally, $LeakyReLU(\cdot, \alpha = 0.2)$ activation function is employed, except in the output layer where we use a $Tanh(\cdot)$ activation, to constraint the output in the range $[-1, 1]$.

The discriminator structure is composed of strided convolutions on which spectral normalization is used, as suggested in I. Gulrajani *et al.* (2017). Dropout layers are triggered after each layer to regularise the discriminator, except for the first

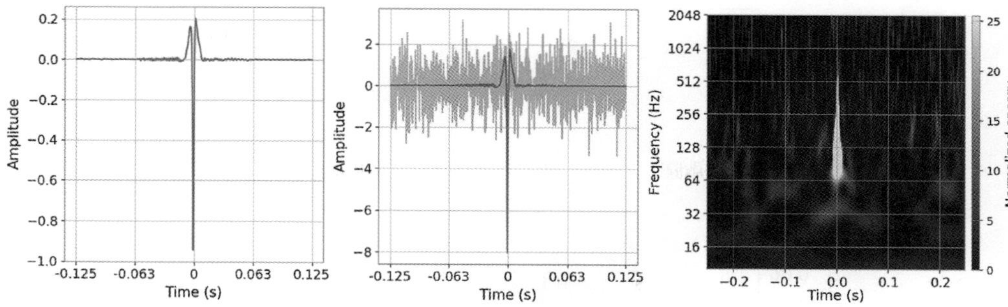

Figure 3. *(Left)* Raw blip as a function of time. *(Middle)* Raw blip injected in real (whitened) noise. *(Right)* Q-transform representation of injected glitch. Note the resemblance between the "fake" glitch here and the real glitch in Fig. 1.

and last layers. The kernel size is set to 5 for all layers, with no padding and $LeakyReLU(\cdot, \alpha = 0.2)$ activation.

During the training, both the generator and the discriminator need to be updated at similar rates for stability and convergence. However, as the classification task is more challenging, we update the discriminator 5 times per generator update. We employ RMSProp optimizer with a learning rate $= 10^{-4}$ for both discriminator and generator, and we train it for 500 epochs. Moreover, we employed $\lambda_1 = 5$, $\lambda_2 = 5$, and dropout rate of 0.6 for convergence. More details about the architecture and the training can be found in M. Lopez *et al.* (2022).

4. Implemented features

Once the model is trained, it is straightforward to use the generator network to produce random glitches. The output of the generator, a raw glitch, is the excess power of a *whitened* time series evaluated on a fixed length time grid at a constant sampling rate 4096 Hz and an amplitude in the range $[-1, 1]$. We can generate a raw glitch in ~ 10 ms on a laptop. A raw glitch can be straightforwardly injected into any whitened time series. In fig. 3 we plot an example of a raw glitch and of how it can be injected in white noise.

Since the amplitude of the generated glitches is arbitrary, before injecting the glitch into noise, a glitch has to be scaled to a user-given Signal-To-Noise (SNR) ratio ρ. The SNR of a glitch $g(t)$ is defined as:

$$\rho^2 = 4 \int_{f_{\min}}^{f_{\max}} \mathrm{d}f \, \frac{|\tilde{g}(f)|^2}{S_n(f)} \tag{6}$$

where again $\tilde{g}(f)$ is the glitch in the frequency domain and $S_n(f)$ is the actual PSD of the (whitened) data where the glitch is injected. The raw glitch is then scaled to achieve the target SNR ρ_{target} using: $g(t)' = \frac{\rho_{target}}{\rho} g(t)$

The raw glitch can also be naturally *resampled*§ at the desired sampling rate. All such functionalities are implemented in the function `get_raw_glitch`: the interested user can find below an example of some working code.

```
import gengli

g = gengli.glitch_generator('H1')
glitch = g.get_raw_glitch(snr = 10, srate = 2048.)
```

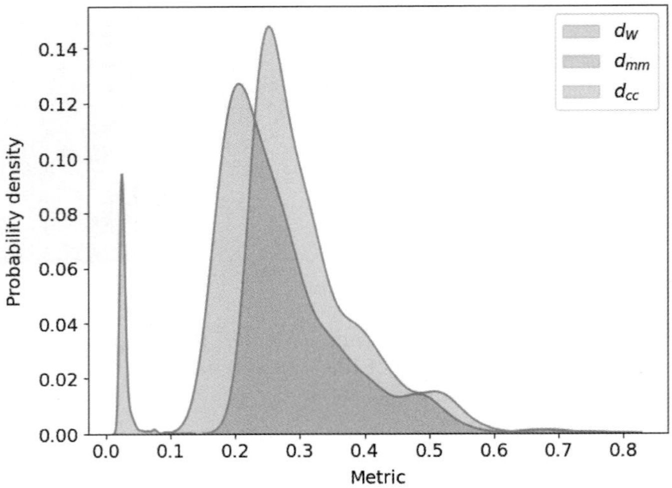

Figure 4. Probability distribution of distances (d_W, d_{mm}, d_{cc}) for a benchmark population of 10^3.

4.1. *Selecting generations*

In M. Lopez *et al.* (2022), it was shown that the generated glitches have the same statistical properties as the training population. However, the training data may (and most likely do) contain anomalous glitches that our model has also learnt to generate (see Section IV of M. Lopez *et al.* (2022)). In this work, we propose a novel method to filter anomalous glitches.

To make this intuition mathematically precise, we start by considering three different measures of distances between an arbitrary pair of glitches:

- Wasserstein distance d_W: a standard measure of distance between distributions, commonly employed in ML.
- Mis-match d_{mm}: a measure of distance based on the details of the filtering, standard in the GW field.
- Cross covariance d_{cc}: we employ the quantity $1-k$, where k is the normalized cross-covariance as defined in M. Lopez *et al.* (2022).

We then populate a benchmark set of N_b glitches with samples from the generator. For each of the $\frac{N_b(N_b-1)}{2}$ pairs of glitches in the benchmark set, we compute the three distances above. In Fig. 4 we show the distribution for the three distances for a population of $N_b = 1000$.

For each new glitch being generated, we compute the set of average distances (d_W, d_{mm}, d_{cc}) between the glitch and the benchmark set, and we measure the set of percentiles (p_W, p_{mm}, d_{cc}) against the benchmark distances. The triplet (p_W, p_{mm}, d_{cc}) is our novelty measure for the glitch that will be used to filter glitches based on an anomaly score interval $[p_{min}, p_{max}]$. The code will output only glitches for which *all* the three anomaly scores lie within the interval.

5. Conclusions

We develop a methodology to generate artificial time-domain blip glitches data with the GAN algorithm. Because of the heavy pre-processing required to deal with real glitches (i.e. denoising, reconstructing etc...), this is a valid way to avoid using real data in state-of-the-art applications. Due to the instability of GAN algorithms, in this particular research, we trained the model constructed in I. Gulrajani *et al.* (2017), modified for time

§ When upsampling, we make the key assumption that there are no interesting features at frequencies higher than 4096 Hz.

series, with the modified Wasserstein loss proposed in X. Wei *et al.* (2018). This model has shown better performance in both training stability and accuracy.

The performance of the generated blips has been assessed in the companion paper M. Lopez *et al.* (2022), where we employ several similarity distances: Wasserstein distance (d_W), mismatch (d_{mm}) and cross-covariance (d_{cc}). The results of these metrics indicate that our model was able to learn the underlying distribution of blip glitches despite the presence of anomalies due to imperfections of the input data set.

In this follow-up work, we introduce our open-source package `gengli`: it provides an easy-to-use interface to the trained GAN output, and has some additional features such as *resampling* and/or *scaling* a glitch and building a population with a given degree of "anomaly".

Future work will condition GAN to other types of glitches (such as *koi-fish* and *tomte*), by using a wealth of data from O3. Our work will enable the GW community to improve glitch classification with ML, study the properties of the glitch population and develop more realistic Mock Data challenges for studies on future GW detectors.

Acknowledgment

V.B. is supported by the Gravitational Wave Science (GWAS) grant funded by the French Community of Belgium, and M.L., S.C and S.S. are supported by the research program of the Netherlands Organisation for Scientific Research (NWO). Computational resources were provided by the LIGO Laboratory and supported by the National Science Foundation Grants No. PHY-0757058 and No. PHY-0823459.

References

J. Aasi *et al.* [LIGO Scientific], Class. Quant. Grav. **32**, 074001 (2015)
B. P. Abbott *et al.* [LIGO Scientific and Virgo], Class. Quant. Grav. **33**, no.13, 134001 (2016)
B. P. Abbott *et al.* [LIGO Scientific and Virgo] Phys. Rev. Lett. **116** no.6, 061102 (2016)
B. P. Abbott *et al.* [LIGO Scientific and Virgo], Phys. Rev. Lett. **119**, no.14, 141101 (2017)
B. P. Abbott *et al.* [LIGO Scientific and Virgo], Astrophys. J. Lett. **892**, no.1, L3 (2020)
R. Abbott *et al.* [LIGO Scientific, VIRGO and KAGRA], [arXiv:2111.03606 [gr-qc]].
F. Acernese *et al.* [VIRGO], Class. Quant. Grav. **32**, no.2, 024001 (2015)
M. Arjovsky, S. Chintala and L. Bottou (2017), arXiv:1701.07875
N. J. Cornish and T. B. Littenberg, Class. Quant. Grav. **32**, no.13, 135012 (2015)
I. J. Goodfellow *et al.*, arXiv:1406.2661.
I. Gulrajani *et al.*, arXiv:1704.00028 (2017)
S. Hild *et al.*, Class. Quant. Grav. **28**, 094013 (2011)
M. Lopez *et al.*, Phys. Rev. D **106**, no.2, 023027 (2022)
M. Maggiore *et al.*, JCAP 03, arXiv:1912.02622 (2020)
A. Torres, A. Marquina, J. A. Font and J. M. Ibáñez, Phys. Rev. D **90**, no.8, 084029 (2014)
X. Wei *et al.* (2018), arXiv:1803.01541
L. Weng, arXiv:1904.08994 (2019)
M. Zevin *et al.*, Class. Quant. Grav. **34**, no.6, 064003 (2017)

Machine Learning in Astronomy: Possibilities and Pitfalls
Proceedings IAU Symposium No. 368, 2025
J. McIver, A. Mahabal & C. Fluke, eds.
doi:10.1017/S1743921322004872

Detecting complex sources in large surveys using an apparent complexity measure

David Parkinson[1] **and Gary Segal**[2,3]

[1]Centre for Theoretical Astrophysics, Korea Astronomy and Space Science Institute, 776 Daedeok-daero, Yuseong-gu, Daejeon 34055, Republic of Korea
email: DavidParkinson@kasi.re.kr

[2]School of Mathematics and Physics, University of Queensland, St Lucia, Brisbane, QLD 4072, Australia

[3]CSIRO Space and Astronomy, PO Box 76, Epping, 1710, NSW, Australia
email: gp.segal@gmail.com

Abstract. Large area astronomical surveys will almost certainly contain new objects of a type that have never been seen before. The detection of 'unknown unknowns' by an algorithm is a difficult problem to solve, as unusual things are often easier for a human to spot than a machine. We use the concept of apparent complexity, previously applied to detect multi-component radio sources, to scan the radio continuum Evolutionary Map of the Universe (EMU) Pilot Survey data for complex and interesting objects in a fully automated and blind manner. Here we describe how the complexity is defined and measured, how we applied it to the Pilot Survey data, and how we calibrated the completeness and purity of these interesting objects using a crowd-sourced 'zoo'. The results are also compared to unexpected and unusual sources already detected in the EMU Pilot Survey, including Odd Radio Circles, that were found by human inspection.

Keywords. Extragalactic radio sources; Astrostatistics; Astrostatistics tools

---

1. Introduction

We are entering an unprecedented era of astronomical data. In the radio wavebands the pathfinders to the Square Kilometer Array will conduct wide and deep surveys. The Australian Square Kilometer Array **Pathfinder** (ASKAP) is predicted to detect ~ 70 million radio galaxies with the Evolutionary map of the Universe (EMU) survey (Norris *et al.* 2011), in comparison to the ~ 2 million extragalactic radio sources currently detected. In the optical, the Legacy Survey of Space and Time (LSST), which will be conducted with the Vera Rubin Observatory (VRO) predicts a detection of 20 billion galaxies and a similar number of stars (Ivezić *et al.* 2019), in comparison with the 300 million galaxies and 80 million stars in the Dark Energy Survey DR1 (Abbott *et al.* 2018), and the 1.6 billion objects in GAIA DR2 (Gaia Collaboration *et al.* 2018).

Many of these objects detected by these new surveys will be of a type never before seen by astrophysicists. This work focuses on a measure of morphological complexity that can be used to identify new and interesting objects. The approach does not requiring learning the specific features of past interesting observations, and as such, can be calibrated on a small sample and then applied to a much larger sample. This method differentiates itself from many existing outlier methods as it does not require the computationally intensive task of computing the distance between specific observations and the rest of the data **in an abstract feature space.** Rather complexity can be computed for **a** single observation and compared to a pre-determined threshold.

2. Methodology

2.1. *Coarse-grained complexity*

We introduce the coarse-grained complexity in Segal *et al.* 2019 based on the notions of *effective complexity* (Gell-Mann 1994, Gell-Mann & Lloyd 1996) and *apparent complexity* (Aaronson, Carroll & Ouellette 2014). The apparent complexity is a measure of the entropy H of an object x computed after applying a smoothing function f, expressed as $H(f(x))$. The Shannon entropy of a probability distribution P can be defined as the expected number of random bits that are required to produce a sample from that distribution:

$$H(P) = -\sum_{x \in X} P(x) \log P(x). \tag{1}$$

As discussed in Aaronson, Carroll & Ouellette 2014, Segal *et al.* 2019, the Kolmogorov complexity $K(f(x))$ can be used as a proxy for the entropy of the smoothed function $H(f(x))$. The Kolmogorov complexity of x is the length of the shortest binary program $l(p)$, for the reference universal prefix Turing machine U, that outputs x; it is denoted as $K(x)$:

$$K(x) = \min_p \{l(p) \ : \ U(p) = x\}. \tag{2}$$

While the Kolmogorov complexity is uncomputable, its upper bound can be reasonably approximated by the compressed file size $C(f(x))$ using a standard compression program such as `gzip`.

Intuitively a complexity measure should provide low values for random data that does not contain structure or regularities. The coarse-grained complexity, like the apparent and effective complexity, can be defined as the compressed description length of regularities and structure after discarding all that is incidental. The coarse-grained complexity achieves this by applying a smoothing function f to the input x (which removes fine-grained noise while preserving the coarse-grained structure of the image) and calibrating this so that complexity values correctly partition data that has been expertly labelled (i.e. by human astronomers).

2.2. *Complexity Scan*

Source detection algorithms may make assumptions, and so miss interesting things. In- stead, we scan the region, computing the course-grained complexity within a sliding frame (square aperture, as shown in Fig. 1).

An important feature of the scan method is that frames are sampled without making any assumptions as to whether the frame contains a conventionally detected source or not (that is without using a source extraction tool or existing catalogue data). This helps reduce the risk of producing a sample that is biased towards preconceived notions of what is interesting, referred to as expectation bias by Norris 2017 and Robinson 1987. This process builds up a 'heat map' of complexity values intended to assist with the identification of the unexpected in new and large data with the goal of new scientific discoveries and surprise.

3. Data

3.1. *EMU Pilot Survey*

The Pilot Survey of the Evolutionary Map of the Universe (EMU-PS) was observed at 944MHz using the Australian Square Kilometre Array Pathfinder (ASKAP) telescope. The ASKAP telescope consists of 36 12-metre antennas spread over a region 6-km in diameter at the Murchison Radio Astronomy Observatory in Western Australia.

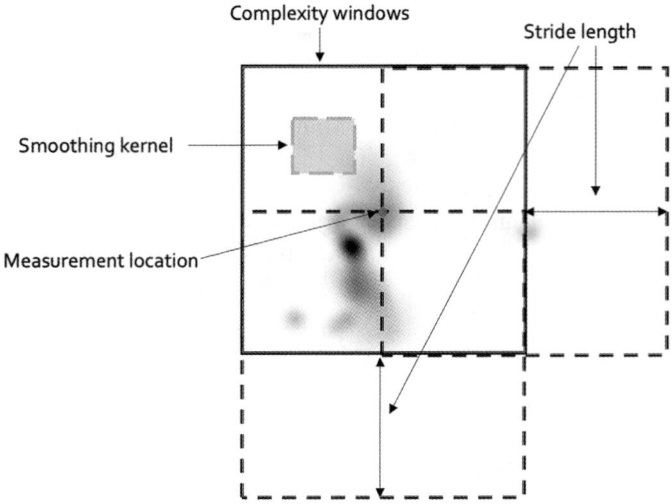

Figure 1. Illustration complexity scan parameters, including: the frame size (side length of solid black square), stride length (offset between position of solid black square and dashed black squares) and smoothing kernel size (grey shaded rectangle).

EMU- PS covers 270 square degrees of an area covered by the Dark Energy Survey at a spatial resolution of $\sim 11 - 13$ arcsec (Norris *et al.* 2021a).

While the primary goal of the EMU Pilot Survey is to test and refine observing parameters and the strategy for the main survey, the pilot in itself presents opportunity for new discoveries. Experience has shown (Norris 2017) that whenever we observe the sky to a significantly greater sensitivitywe make new discoveries. This goal has already been demonstrated through the successful identification of a new class of radio object, odd radio circles (ORCs, Norris *et al.* 2021b, Koribalski *et al.* 2021, Norris *et al.* 2022).

3.2. *Anomaly Zoo*

Truth labels are required to evaluate the effectiveness of alternative complexity thresholds for partitioning anomalous sources. To provide truth labels for the frames produced by the complexity scan we ran a project on the `Zooniverse.org` platform, titled *"Anomaly in the EMU Zoo"* (hereafter zoo), requesting expert astronomers to evaluate an unbiased sample of frames sub-sampled from the EMU-PS scan. Consensus from the zoo labels was then used to evaluate the Recall and Precision associated with prospective partition boundaries.

Expert volunteers were approached from within the Evolutionary Map of the Universe Survey Project and at the SPARCS 2021 conference. A sub-sample of 1627 frames from the EMU-PS scan ($n_{\text{total}} = 365{,}000$) were presented to volunteers for classification through the Zooniverse project. 44 volunteers participated in the project, with 10 of these classifying more than 500 frames each.

The zoo asked the expert volunteers to evaluate frames sub-sampled from the EMU-PS Scan and to select an option that best describes the most interesting radio sources in each frame before moving on to the next. The four options presented for selection were:

- No sources/just noise
- One or more simple sources/unrelated simple sources
- At least one complex source/sources with multiple components
- Contains something unexpected/Anomaly

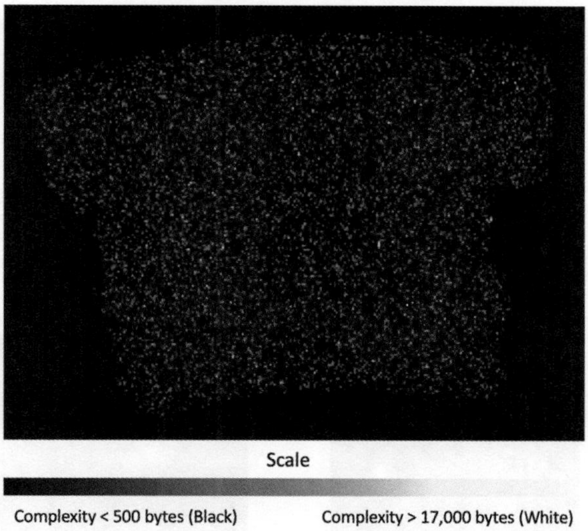

Figure 2. A complexity 'heat map' of the EMU-PS region, where the complexity of the frame is indicated by the brightness of the pixel. While many of the frames are red (grey) coloured, indicating the presence of a simple source, there are a few yellow (light grey) or white pixels, indicating more complex sources or low-surface brightness structures.

The zoo distinguished between complex and extended sources with multiple components, and sources that were deemed by the volunteers to be truly unexpected or anomalous. Only frames converging on a label through majority consensus were used to evaluate the effectiveness of the complexity to identify anomalous sources. Those frames from the zoo sample where majority consensus was not reached were excluded from the evaluation.

4. Results

4.1. *Complexity Heat Map*

We performed a complexity scan of the EMU-PS data using the method described. Com- plexity values are shown in Fig. 2 as a heat map overlaid on the EMU-PS field.

Visual inspection of a sub-sample of frames shows that the high-complexity value tail of the distribution comprises unusual, complex and extended objects with examples shown in Fig. 3. These include peculiar Tailed Radio Galaxies, **Fanaroff-Riley Class I (FRI) and Class II (FRII)** type sources with more inflated lobes than that of average such objects, Odd Radio Circles, and Giant Radio Galaxies including a previously unreported Giant Radio Galaxy **with projected linear length** of 1.94 Mpc.

4.2. *Anomaly Zoo Outcomes*

We use truth labels determined from the zoo to evaluate the effectiveness of alternative partitions for identifying anomalous objects. As an immediate benefit, an effective partition can be used to identify frames containing complex structures and unusual objects from the EMU-PS data and build an anomaly catalogue. An effective boundary can also be used when analysing new, and even larger surveys, including the subsequent full EMU survey which is anticipated to capture approximately 40 million sources.

Through sub-sampling of the EMU-PS Scan, the sample size for the zoo was limited to 1,627 frames. An enrichment sample was included to provide better representation of the type of observations found within frames beyond the 99.5th percentile. Classification

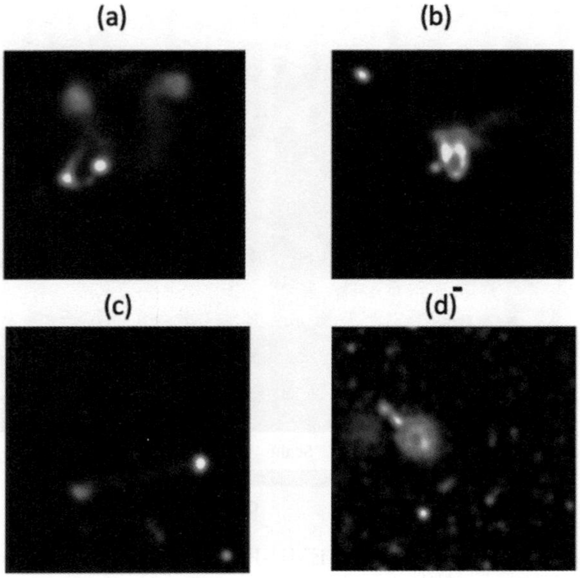

Figure 3. Examples of objects found within frames sampled above the 99th percentile complexity value from the EMU-PS Complexity Scan. These examples illustrate the breadth of objects found within frames in the complexity tail including, (a) the unusual source PKS 2130-538 (Otrupcek & Wright 1991) known as the Dancing Ghosts (Norris *et al.* 2021a), (b) a bright wide angle tail source on 2MASX J21291901-5053040 in cluster Abell 3771, (c) a previously unreported Giant Radio Galaxy **with projected linear length** of 1.94 Mpc (DES J221640.02-625241.6), (d) two odd radio circles, EMU PD J205842.8–573658 (ORC2) and EMU PD J205856.0-573655 (ORC3).

Table 1. Classification counts by consensus for each of the options provided in the EMU-PS Anomaly in the EMU Zoo. The enrichment sample supplements the zoo sub-sample with frames measuring complexity above the 99.5th percentile.

Classification	**Count: Non-bias sample (n = 1528)**	**Count: Enrichment (n = 99)**	**Count: Combined Sample (n = 1627)**
Anomalous	7	17	24
Complex	366	57	423
Simple	1059	12	1071
No source	31	0	31
No consensus	65	13	78

counts for each of the zoo classes are shown in **Tab**. 1. This table includes the anomaly counts both before and after the inclusion of the enrichment sample.

Only frames converging on a label through majority consensus were used to evaluate the effectiveness of the complexity to identify anomalous sources. Those frames from the zoo sample where majority consensus was not reached were excluded from the evaluation. 99.8% of the zoo sample frames received evaluations from 3 or more volunteers and 94% from 5 or more. We used the criterion that only frames receiving 5 or more evaluations were used to evaluate consensus, and be given a reliable 'truth' label. These restrictions were imposed to avoid the results being impacted by outlier evaluations that differed from the majority of expert volunteers.

We partition the data based on complexity and **signal-to-noise ratio (SNR)** values and use this to define a function based catalogue boundary (shown in Fig. 4).

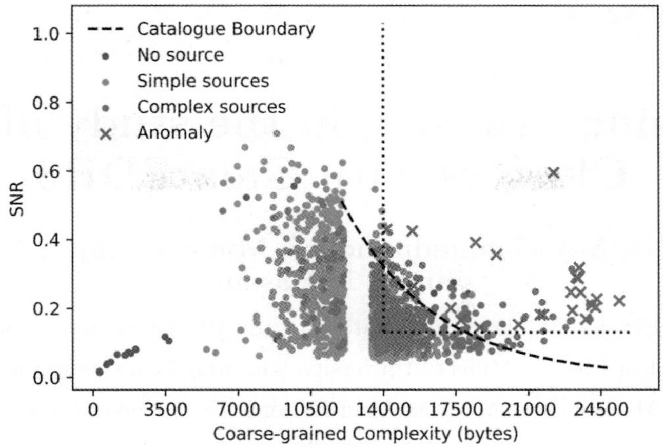

Figure 4. Scatter plot showing the coarse-grained complexity and the signal-to-noise ratio (SNR), and the relationship between them, along with the classification of frames derived through consensus from the zoo (Anomaly in the EMU Zoo).

5. Summary

The coarse-grained complexity can be used as a tool for identifying unusual and complex objects. We apply the method to new EMU-PS data (365,000 sampled frames containing at least the 220,000 Selavy catalogue sources) to identify and segment unusual sources (i.e. anomalies).

We used a Zooinverse project to produce crowd-sourced labels to evaluate the effectiveness of the approach and we identified an effective anomaly partition using the coarse-grained complexity and SNR values.

Results demonstrate the ability of the coarse-grained complexity to single out regions of the sky that contain complex and unusual sources, in a manner that can be computed at worst-case linear time complexity without reference to existing catalogue data. We propose the complexity measure as useful tool for identifying regions of interest in subsequent large and deep radio continuum surveys.

References

Aaronson S., Carroll S., Ouellette L., 2014, `http://arxiv.org/abs/1405.6903v1`

Abbott, T. M. C., Abdalla, F. B., Allam, S., Amara, A., Annis, J., *et al.* 2018, *ApJS* 239, 18

Gaia Collaboration *et al.* 2018, *A&A* 616, A1

Gell-Mann, M., 1994, *"The Quark and the Jaguar: Adventures in the Simple and the Complex,"* Henry Holt and Company

Gell-Mann, M. & Lloyd, S. 1996, *Complexity*, 2, 1

Ivezić, Ž, Kahn, S. M., Tyson, J. A., Abel, B., *et al.* 2019, *ApJ* 873, 2

Koribalski, B. S., Norris, R. P., Andernach, H., Rudnick, L., Shabala, S., Filipović, M., Lenc, E. 2021, *MNRAS Letters*, 505, 1

Norris, R. P., Hopkins, A. M., Afonso, J., Zinner, *et al.* 2011, *PASA*, 28, 3

Norris, R. P. 2017, *PASA*, 34, e007

Norris, R. P. *et al.* 2021a, *PASA*, 38, e046

Norris, R. P. *et al.* 2021b, *PASA*, 38, e0003

Norris, R. P. *et al.* 2022, *MNRAS*, 513, 1

Otrupcek, R. E. & Wright, A. E. 1991, *PASA*, 9, 1

Robinson, B. J. 1987, *PASA*, 7, 220

Segal, G., Parkinson, D., Norris, R. P., Swan, J., 2019, *PASP* 131, 1004

Machine Learning in Astronomy: Possibilities and Pitfalls
Proceedings IAU Symposium No. 368, 2025
J. McIver, A. Mahabal & C. Fluke, eds.
doi:10.1017/S1743921323000807

Machine Learning in the study of Star Clusters with Gaia EDR3

Priya Hasan[1], Md Mahmudunnobe[2], Mudasir Raja[1], Md Saifuddin[1] and S N Hasan[3]

[1]Department of Physics, Maulana Azad National Urdu University, Hyderabad 500032, India

[2]Department of Physics, Minerva University, San Francisco, California 94103, USA

[3]Department of Mathematics, Maulana Azad National Urdu University, Hyderabad 500032, India,
email: priya.hasan@gmail.com

Abstract. Determination of membership of star clusters is a very important criterion in their study as they effect determination of cluster parameters like radius, age, distance, mass functions, etc. In an earlier study, we used published membership data of nine open star clusters as a training set to find new members from Gaia DR2 data using a supervised Random Forest (RF) model with a precision of around 90%. The number of new members found was almost double the published number. In this work, we would like to compare the earlier results with results obtained by applying the unsupervised method of Gaussian Mixture Modelling (GMM) and Density-Based Spatial Clustering of Applications with Noise (DBSCAN) to study the membership of open star clusters of varying ages and locations in the Galaxy using Gaia DR2 and EDR3 data. We shall discuss these techniques and focus on the caveats involved.

Keywords. Machine Learning, GMM, Gaia EDR3, Star Clusters, Membership

1. Introduction

Star clusters are groups of stars born from the same molecular cloud, at the same distance, velocity as well as of the same chemical composition. They are the ideal stellar laboratories to study stellar formation, evolution and dynamics. They are the basic building blocks of galaxies and to understand them, the most crucial parameter is membership.

The unprecedented accuracy of Gaia data has lead to a paradigm shift in open cluster membership determination, and numerous machine learning (ML) models discussed in Hunt & Reffert (2021) are used to accomplish this. This is based on the premise that stars of a cluster should appear 'clustered' in an n-dimensional space made up of their coordinates and velocities. By increasing the number of members, we can study the outer regions of clusters, revise parameters like radius, age, distance, mass functions, etc. and enrich our color magnitude diagrams (CMDs). Various ML methods, by Cantat-Gaudin & Anders (2020), Agarwal et al. (2021); Gao et al. (2020); Mahmudunnobe et al. (2021) have been used for membership determination of star clusters using ML.

Various methods allow for a range of approaches to the cluster identification problem, such as centroid-based algorithms (k-means algorithm), distribution-based clustering (such as Gaussian-mixture models, GMM), or density-based algorithms. (For a summary of clustering analysis in astronomy, see Feigelson & Babu (2012), Chap. 3.3 and the references therein.). Some methods are useful in locating clusters with arbitrary shapes

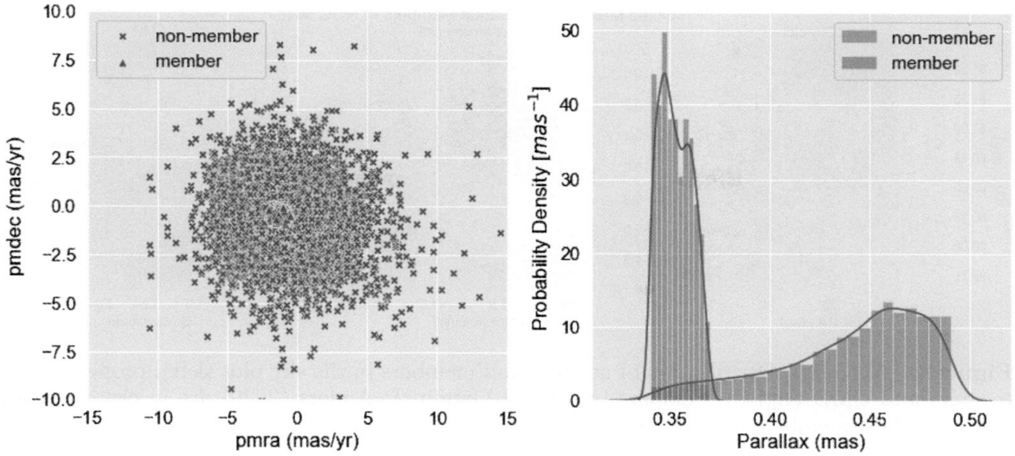

Figure 1. NGC 581: Separation of cluster and field stars in proper motion (left) and parallax (right) space. Members identifies by GMM are in green triangles and field stars are in blue crosses.

that can be generically characterised as overdensities in a low density environment. The advantage of density-based algorithms is that they don't assume anything about the dataset being studied beforehand. In other words, the user does not need to be aware of the number of clusters present in the dataset because these algorithms do not assume any distribution while associating the data points with a cluster.

In an earlier paper, we applied the supervised method of Random Forest (RF) and used published membership data of nine open star clusters as a training set to find new members from Gaia DR2 data using a supervised RF model with a precision of around 90%. The number of new members doubled the number in the training sets, in most cases Mahmudunnobe et al. (2021). In this paper, we compare our results using the GMM and Density-Based Spatial Clustering of Applications with Noise (DBSCAN) (Ester et al. 1996) methods using Pedregosa et al. (2011).

About 1.8 billion sources are catalogued in Gaia EDR3 (Riello et al. 2021) and references therein with data on parallaxes, proper motions (*pmra*, *pmdec*, and the (G, G_{BP}, G_{RP}) magnitudes. This work will involve use of this dataset.

2. Gaussian Mixture Modelling (GMM)

GMM is a parametric ML model. The primary assumption behind GMM is that the observed dataset is generated from a mixture of two or more Gaussian (i.e. normal) distributions with unknown distribution parameters. The most important input parameter of GMM is the number of components, k (i.e. number of Gaussian distributions that generated the data). We filtered the Gaia ER3 data for 9 clusters namely, M67 (NGC 2682), NGC 6791, NGC 2243, NGC 2141, NGC 6823, NGC 6231, IC 1805, NGC 1893 and NGC 581

Figure 1 shows the results for NGC 581 with a clear separation of cluster and field stars in proper motion (left) and parallax (right) analysis for NGC 581. Figure 2 shows for NGC 581, the distribution of newly found members in the sky plot (left) proper motion (center) and CMD (right). In both figures members by Cantat-Gaudin & Anders (2020) are in orange crosses and our GMM members are in green triangles.

In this study, we define a measurable statistic called the Modified Silhouette Score (MSS) to assess how well unsupervised models do in determining membership. We examine the effectiveness of the GMM to identify members from the Gaia DR3 data.

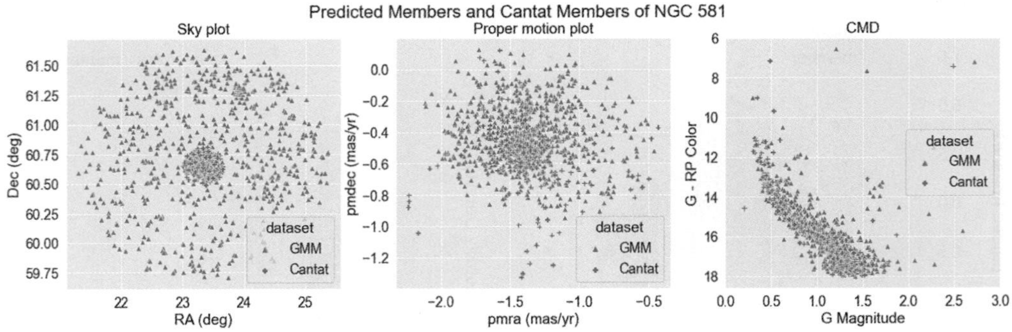

Figure 2. NGC 581: Distribution of newly found members in the sky plot (left), proper motion (center) and CMD (right). Members by Cantat-Gaudin & Anders (2020) are in orange pluses and our GMM members are in green triangles.

We investigate the dependence of MSS on age, distance, extinction, galactic latitude and longitude, and other characteristics. We discover that GMM works best for clusters that are closer and relaxed.

MSS can be used to measure the performance of an unsupervised model, which includes two sets of stars: members and field stars. For star cluster membership determination, we assume a normal distribution with a narrow standard deviation (SD) for members and a distribution with broad SD for field stars in the feature space with k features.

$$MSS = \frac{1}{K}\sum_{i=1}^{K}\frac{SD_{member} - SD_{field}}{max(SD_{member}, SD_{field})}$$

MSS can range from -1 to 1. The higher the MSS value, the better the performance of the model to distinguish members from field stars.

3. Density-Based Spatial Clustering of Applications with Noise (DBSCAN)

DBSCAN is an unsupervised density-based clustering method that is well suited for open clusters since it can find clusters of non-spherical shapes. The average density of stars in the area is required by the DBSCAN algorithm; in smaller regions, this average is more representative than if we take the entire sky, where the density can vary widely from one place to another. We first normalise the data before using DBSCAN to make sure that the variable range differences do not affect the analysis. Around the cluster centre, we selected a search radius of 0.5 degrees for far-off clusters and 1 degree for nearby clusters. DBSCAN groups stars into clusters and checks the minimum number of stars, $mPts$ within an ϵ radius of stars. If a minimum of $mPts$ points are found in the ϵ radius of a point, it is called a core point. Points which lie in the ϵ radius of a core point but do not have minimum $mPts$ are called border points and points outside the ϵ radius which do not have minimum $mPts$ are noise points. The DBSCAN group with the maximum number of stars is selected as the cluster member group.

We studied the membership of nine open clusters (NGC 581, NGC 1893, IC 1805, NGC 6231, NGC 6823, NGC 3293, NGC 6913, NGC 2264, NGC 2244) using the DBSCAN clustering algorithm on Gaia DR3 Data. We select stars from the Gaia DR3 catalogue, construct a five-dimensional phase space (three-dimensional spatial position and two-dimensional proper motion) and obtain reliable cluster members using DBSCAN as described earlier.

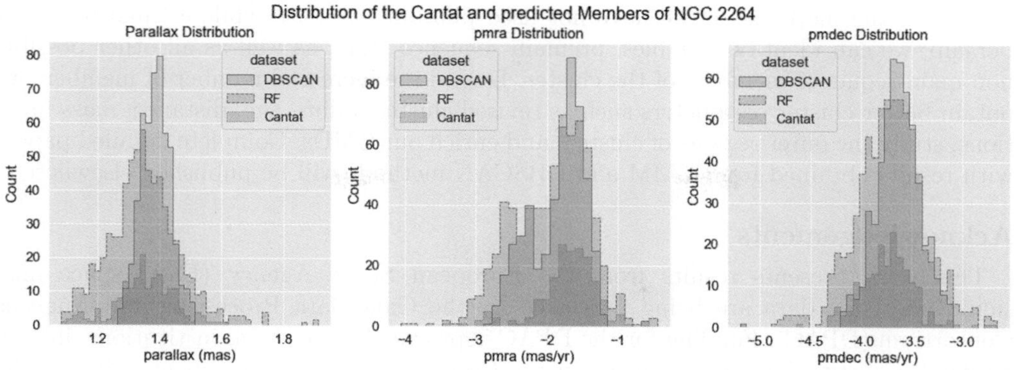

Figure 3. NGC 2264 cluster:Parameter distribution. Parallax (left), *pmra* (center) and *pmdec* (right). Stars from our method are in green solid line, Mahmudunnobe et al. (2021) in purple dashed line and Cantat-Gaudin & Anders (2020) in orange dotted line.

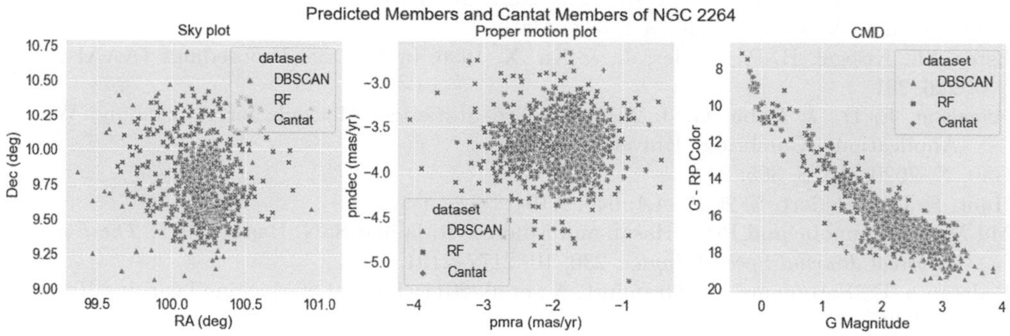

Figure 4. Predicted the members in a sky plot (left), proper motion (center) and CMD (right. Stars from our method are in green triangles, Mahmudunnobe et al. (2021) in purple crosses and Cantat-Gaudin & Anders (2020) in orange pluses).

We applied our method to NGC 2264 and show the results in Fig. 3. We show the distribution of parameters: parallax (left), *pmra* (center) and *pmdec* (right) obtained using our method. Figure 4 shows the distribution of members in a sky plot (left), proper motion (center) and CMD (right) where our new members are green triangles, Mahmudunnobe et al. (2021) are orange pluses and Cantat-Gaudin & Anders (2020) are purple crosses. There is a good agreement and the number of members has been substantially increased.

4. Conclusions

The performance of the GMM method for membership determination can be evaluated using the MSS metric. For GMM, we compared nine clusters with varying ages but we did not find any significant differences between GMM performance for younger and older clusters. But we found a moderate correlation between GMM performance and the cluster distance, where GMM works better for closer clusters. There is little to no link between any other variables and GMM performance.

We used DBSCAN method to find the membership of nine open clusters with Gaia DR3 data. We found that the number of member stars in clusters was increased by a large factor, almost four times.

In both the methods, as we have not used photometric data while estimating membership, we can identify variables, premain sequence stars, as well as all other possible non main-sequence members of the cluster. Due to the increased number of members we obtain better cluster parameters such as revised cluster radius, age, distance, mass functions, study the outer regions of clusters and enrich our CMDs. Complete detailed papers with results obtained from GMM and DBSCAN methods will be published elsewhere.

Acknowledgements

This work presents results from the European Space Agency (ESA) space mission Gaia. Gaia data are being processed by the Gaia Data Processing and Analysis Consortium (DPAC). Funding for the DPAC is provided by national institutions, in particular the institutions participating in the Gaia MultiLateral Agreement (MLA). The Gaia mission website is https://www.cosmos.esa.int/gaia. The Gaia archive website is https://archives.esac.esa.int/gaia.

References

Agarwal, M., Rao, K. K., Vaidya, K., et al. 2021, *MNRAS*, 502(2), 2582–2599

Cantat-Gaudin & Anders, 2020, *AA*, 633, A99

Ester, M., Kriegel, H.-P., Sander, J., & Xu, X. 1996, in KDD-96 Proceedings (AAAI Press), 226–231

Feigelson, E. D., & Babu, G. J. 2012, Modern Statistical Methods for Astronomy: With R Applications (Cambridge University Press).

Gao, X. 2020, *ApSS*, 365, 24.

Hunt, E. L. & Reffert, S. 2021, *AA*, 646, A104.

Md Mahmudunnobe and Priya Hasan and Mudasir Raja and S. N. Hasan, 2021, *The European Physical Journal Special Topics*, 230, 10, 2177–2191

Pedregosa, F., Varoquaux, G., Gramfort, A., et al. 2011, Journal of Machine Learning Research, 12, 282

Riello, M., De Angeli, F., Evans, D. W., 2021, *AA*, 649, A3.

Machine Learning in Astronomy: Possibilities and Pitfalls
Proceedings IAU Symposium No. 368, 2025
J. McIver, A. Mahabal & C. Fluke, eds.
doi:10.1017/S1743921322003568

Assessing the Quality of Massive Spectroscopic Surveys with Unsupervised Machine Learning

John F. Suárez-Pérez and Jaime Forero-Romero

Departamento de Física, Universidad de los Andes, Cra. 1 No. 18A-10, Bogotá, Colombia
emails: `jf.suarez@uniandes.edu.co`, `je.forero@uniandes.edu.co`

Abstract. Massive spectroscopic surveys targeting tens of millions of galaxies are starting to dominate the observational landscape in the 2020 decade. For instance, a night of observation with the Dark Energy Spectroscopic Instrument (DESI) can measure around of 100,000 spectra, with each spectrum sampled over 2,000 wavelength points. Assessing the quality of such a massive data flow requires new approaches to complement visual inspection by humans. In this work, we explore the Uniform Manifold Approximation and Projection (UMAP) as a technique to assess the data quality of DESI. We use UMAP to project DESI data into a 2-dimensional space. In this space, we are able to find outliers that correspond to instrument fluctuations that can be fully diagnosed by inspecting the raw data. These results pave the way for to use machine learning to monitor the health of massive spectroscopic surveys automatically.

Keywords. methods: data analysis, techniques: miscellaneous, surveys

1. Introduction

The Dark Energy Spectroscopic Instrument (DESI) is a large spectroscopic survey that started observation operations in 2020 and will measure spectra for 5 years. DESI will observe around 40 million spectra. These observations include three types of objects: Galaxies, Quasars, and Stars. On an observation night, DESI can measure on the order of 100,000 spectra, with each spectrum sampled over 2,000 wavelength points (DESI Collaboration et al. (2016)). Examining and detecting possible instrumental errors is a difficult task due the massive information. Complete inspection is beyond human capabilities.

In the new astronomical era, where the amount of observations is growing, some strategies have found support from artificial intelligence tools to organize, analyze, process and understand the data (Way et al. (2012)). For these massive astronomical observations, the quality of the observations is fundamental. For this reason, monitoring the quality of this massive data with new methods is a task that improves the data acquisition process. New methods based on machine learning have demonstrated good results in identifying abnormal objects in massive datasets (Baron and Poznanski (2017)).

Following a similar methodology, we proposed a machine learning-based method composed of two mechanisms. The first one, is a dimensionality reduction algorithm, an unsupervised machine learning algorithm that projects high dimensional data into a reduced space that eases the inspection. Then, an algorithm identifies the outliers in the low-dimensional representation. Using this method we found that the outliers identified in a set of DESI observations are related to instrumental errors.

2. Methodology

In an observation process, the acquisition and processing steps are open to instrumental or analysis errors. To identify these errors in massive spectral observations, we proposed a learning-based method. This method allows us to identify the instrumental errors as outliers in reduced representation space.

Our method consists of two algorithms: First, the Unsupervised Manifold Approximation and Projection (UMAP)(McInnes et al. (2018)), a dimensionality reduction algorithm that projects the spectra in a two-dimensional space. The second, the Friend of Friends (FoF) algorithms that identify the outliers in the reduced space (Huchra & Geller (1982)).

2.1. *UMAP*

UMAP is an algorithm that uses a high dimensional graph representation of the data to find a low-dimensional graph representation preserving the manifold structure of the input data (McInnes et al. (2018)). This means that similar objects in the input space will be allocated to locations nearby in the reduced space.

To find this reduced graph representation UMAP has some meta parameters that can be tuned. The *metric* and the *number of neighbors* are meta-parameters that work on the input space. These meta-parameters are used to build the representation graph. The *minimal distance* and the *dimension* allow for different representations on the output space.

• **Metric:** This meta-parameter defines the mechanism to measure the distance between objects on the input space. In this case we explore between the *cosine*, *euclidean* and *braycurtis* metrics.

• **Number of neighbors:** This meta-parameter defines the connections between the objects in the input space. A low number of neighbors will focus the algorithm on the local structure, whereas, if the number of neighbors is high the algorithm will focus on the global structure, losing the fine structure. For this meta-parameter, we explore values from 0 to 50 with a step of 5.

• **Minimal Distance:** The minimal distance defines how near the objects will be projected on the output space. A low minimal distance will produce clusters, while, a high value in the *minimal distance* will produce a sparse representation. For this meta-parameter, we explore values from 0.0 to 0.5 in steps of 0.1.

• **Dimension:** This meta-parameter defines the dimension of the output space. In our work, we fixed it to a 2-dimensional for visualization purposes.

2.2. *FoF*

The FoF algorithm is used to identify groups in feature space (Huchra & Geller (1982)). An object belongs to a group if it is located in a spherical region with a radius defined as a *linking length.* The FoF algorithm has a unique meta-parameter that controls the connection length to identify friends. For this meta-parameter, we explore distances from 0.1 to 0.5 in steps of 0.1.

3. Preliminary Results

To evaluate the efficiency of our method we use the data from the first DESI Data Release (DR), Andes. Andes was the first set of calibrated observations and includes around 100,000 spectra, each one with 2,000 points by band, B, R, and Z, and 6,000 considering the full spectral bandwidth from 3000 Å to 9600 Å.

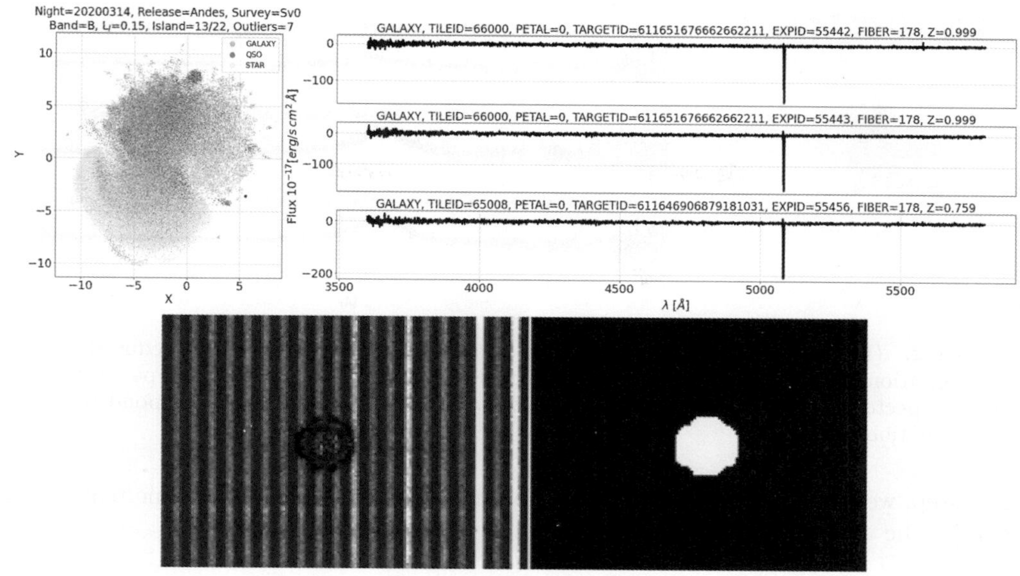

Figure 1. Example of an instrumental error identified on the 14 March 2020 night from the Andes DR. This error was identified as a defect in the CCD camera. (Top) The left box shows the 2-dimensional representation obtained with UMAP and a set of outliers highlighted with black. The right box shows some spectra examples with an abnormal feature and large negative flux. (Bottom) This plot shows the spectrum projected on the CCD and the burned pixels.

3.1. *Instrumental error*

Figure 1 shows the example of our method applied on the night of 14th March 2020. To obtain this projection we use a *cosine* metric, with 15 *nearest neighbors*, a *minimal distance* of 0.5 and a *linking length* of 0.2. The left box shows the UMAP projection for the observations in the B Band. The colors correspond to four classes, green for the stars, blue for the galaxies (except Bright Galaxies, BG), magenta for the QSO, and red for BG. A group of outliers is highlighted in black. The right box shows three examples of spectra extracted from the highlighted sample. These spectra share the same rare feature, large negative flux around 5100 Å. This feature could mean a calibration error.

The identification of these abnormal spectra was reported to the DESI spectral data group. After visual inspection, it was identified that the error corresponds to an instrumental error, a defect in the CCD camera†, Figure 1 bottom.

3.2. *Contamination on the observations*

Using our method we identify another kind of error. The collaboration reported abnormal spectra on the project's GitHub repository‡. Figure 2 (central) shows an abnormal spectrum. Our method identified these rare spectra projected in the reduced space as an outlier (left) using the cosine metric, 15 nearest neighbors, 0.5 as minimal distance, and a linking length of 0.2. In this outlier were identified 163 objects in total. All of these have the same shape. The DESI spectral data group identified this feature as contamination on the observation, Figure 2 (right).

In future investigations, we will inspect the 1% of the DESI observations to test our method. This 1% will be the first public Data Release of the DESI experiment. In this

† `https://github.com/desihub/desispec/issues/983`
‡ `https://github.com/desihub/desispec/issues/1262`

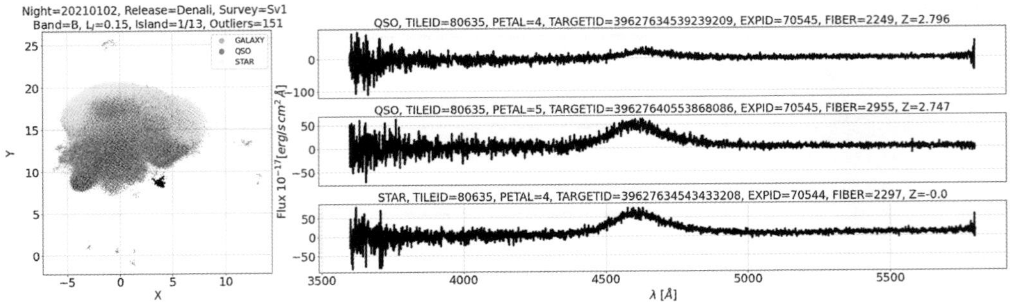

Figure 2. (Left) Highlighted in black, abnormal spectra identified on the reduced UMAP representation. (Right) Spectra sample that evidence an unusual spectral shape. A total of 151 unexpected spectra were detected for that night. That result might correspond to stellar contamination.

next step, we will estimate the use of memory resources and computing time required to identify the outliers required by our method.

4. Summary

We have presented a method learning-based for the detection of instrumental errors in spectroscopic surveys. Our method is novel and useful to assess the quality of massive astronomical surveys. The method consists of two algorithms: The Uniform Manifold Approximation and Projection (UMAP) an unsupervised dimensionality reducer machine learning algorithm, that project high-dimensional data into 2-dimensional representations.

Then, a Friend-of-Friends (FoF) algorithm identifies the outliers groups that contain abnormal spectra. In this work, we use the spectra from surveys validation of the Dark Energy Spectroscopic Instrument (DESI). Each DESI spectrum, with around 2,000 points by band (B, R, or Z) or 6,000 for concatenated spectra is projected in a 2-dimensional embedding space. Visual inspection of the embedding reveals outliers that correspond either to CCD fluctuations or stellar contamination.

Our results show a promising method to identify instrumental errors on massive spectroscopic surveys. This method is necessary now that surveys like DESI2, EUCLID, LSST, WFIRST, and the CSST (Chinese Space Station Telescope) will have a large amount of data.

5. Acknowledgements

JFSP and JEFR acknowledge the support of the INV-2022-133-2337 and INV-2022-137-2394 projects of the Universidad de Los Andes, Facultad de Ciencias.

References

Baron, D., Poznanski, D. 2017. The weirdest SDSS galaxies: results from an outlier detection algorithm. *MNRAS* 465, 4530–4555. doi:10.1093/mnras/stw3021

DESI Collaboration, Aghamousa, A., Aguilar, J., et al. 2016, The DESI Experiment Part I: Science,Targeting, and Survey Design, arXiv:1611.00036

Huchra, J.P. & Geller, M. J. 1982, Groups of Galaxies. I. Nearby groups, *ApJ*, 257, 423. doi:10.1086/160000

McInnes, L., Healy, J., & Melville, J. 2018, UMAP: Uniform Manifold Approximation and Projection for Dimension Reduction. arXiv:1802.03426

Way, M. J., Scargle, J. D., Ali, K. M., Srivastava, A. N. 2012. Advances in Machine Learning and Data Mining for Astronomy.

Machine Learning in Astronomy: Possibilities and Pitfalls
Proceedings IAU Symposium No. 368, 2025
J. McIver, A. Mahabal & C. Fluke, eds.
doi:10.1017/S1743921323000704

Neural Networks for Meteorite and Meteor Recognition

Aisha Al-Owais, Maryam Sharif, Ilias Fernini and Antonios Manousakis

Sharjah Academy for Astronomy, Space Sciences, and Technology
email: aalowais@sharjah.ac.ae

Abstract. In this paper, we present a convolutional neural network (CNN)-based architecture trained on a dataset of meteorites and terrestrial rocks and another dataset trained on meteors and light sources. For meteorites, the dataset comprises augmented images from the meteorite collection at the Sharjah Academy for Astronomy, Space Sciences, and Technology (SAASST). For meteors, the images are taken from the United Arab Emirates (UAE) Meteor Monitoring Network (MMN). Such a project's significance is to expand machine learning applications in astronomy to include the solar system's small bodies upon contact with the Earth's atmosphere. This, in return, acts as deep learning research, which examines a computer's ability to mimic a human's brain in recognizing meteorites from rocks, and meteors from airplanes and other noise sources. When testing the CNN models, results have shown that both the meteorite and meteor models reached an accuracy of above 80%.

Keywords. Meteors, Meteorites, Neural Networks, Machine Learning

1. Introduction

Distinguishing meteorites from rocks can be very challenging due to the similar features they possess. Recovering them from desert areas with high temperatures, like the United Arab Emirates (UAE), puts meteorite hunters' health at risk in addition to consuming a considerable amount of time. Attempts to integrate deep learning approaches with drones for meteorite recuperation have been explored by many, most recently by (Citron et al. 2021) and (Anderson et al. 2020). The latter, were able to recover an actual meteorite in Australia (Anderson et al. 2022). Meteors have also become a target for image classification and object detection, especially with the meteor monitoring networks increasing globally such as (Vida et al. 2021), and many others. In (Gural 2019) neural networks were utilized to detect meteors, reaching remarkable accuracy levels. The gap remains in utilizing the You Only Look Once (YOLO) neural network to detect meteorites as well as meteors, which this paper attempts to fill.

2. Methodology

YOLOv3, a Convolutional Neural Network (CNN)-based architecture (Redmon and Farhadi 2018), was trained on our meteorite collection and images of rocks. While other models were developed for the same purpose, their models were not targeted to distinguishing meteorites from rocks. Each class contained 380 images of original meteorite and rock samples, and 3,036 images post data augmentation. The objects were manually labelled using a labeling tool (Cartucho et al. 2018) and were then trained using YOLOv3. It was then deployed on a Jetson Nano which utilized an RGB camera, carried by a drone, for on-site detection. The model is also used for off-site detection. The same

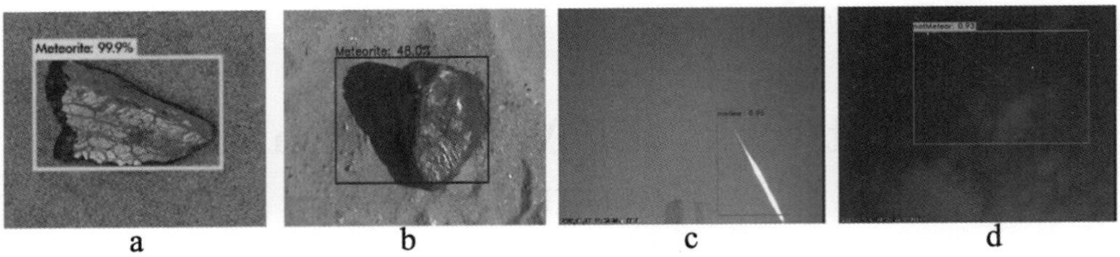

Figure 1. A high-resolution image of a meteorite is tested in (a) where the model gave an accuracy rate of 99%. However, on-board testing showed a much lower rate (b). A meteor is correctly classified with a rate of 96% (c) and a not-meteor is classified as such with a rate of 93% (d).

object detector was also trained on a data set of meteors and false meteors obtained from the UAE Meteor Monitoring Network (MMN) (Fernini et al. 2020). The network observes objects in the UAE sky daily from sunset to sunrise, and is comprised of three towers, each having 17 cameras. Each class comprised 4,000 images after data augmentation. After training, the model is used to detect objects in the newly captured images. Since bounding boxes appear on each object, either a meteor or not, it becomes relatively faster for the individual to filter out the true detections from the false ones.

3. Results

For meteorites, the model provided better accuracy rates when tested on high resolution images compared to images from the Raspberry Pi 3 camera. Thus, testing in real-time produces lower results than testing images off-board (Figure 1). As for meteors, the model performed remarkably well in detecting meteors as well as not-meteors. Nevertheless, it also has a relatively high number of false positives, especially if the object was an airplane. Since YOLOv3 utilizes the value of the mean Average Precision (mAP) as a metric to evaluate a model's performance, the obtained mAP value for the meteorite model was 97% and for meteors 68%. Nevertheless, accuracy rates for both models when tested on the desired object, reached more than 80%.

4. Conclusion

We have presented an application of YOLOv3 on meteorites and meteors. The meteorite model performed well on post-processed images but showed a lower accuracy percentage when tested on-board due to the camera's low resolution. This could be enhanced by feeding the CNN with images taken from a RPi3 camera as well as drone images taken from varying altitudes. Coming to meteors, the model was able to detect meteors, including faint ones. Yet, it still yielded false positives that looked like meteors. The meteor model is currently implemented with the UAEMMN system and reduces data, while the meteorite model is being deployed on a single processing board along with an UAV.

Supplementary material

To view supplementary material for this article, please visit https://doi.org/10.1017/S1743921323000704

References

Citron, R., *et al.* 2021, MAPS, 56, 6, 1073–1085
Anderson, S., *et al.* 2021, MAPS, 55, 11, 2461–2471
Anderson, S., *et al.* 2021, ApJL 930, 2

Vida, D., *et al.* 2021, MNRAS, 506, 4, 5046–5074
Gural, S. 2019, MNRAS, 489, 4, 5109–5118
Redmon, J. and Farhadi, A. 2018, arXiv preprint arXiv:1804.02767
Cartucho, J., *et al.* 2018, 2018 IEEE/RSJ IROS, 2336–2341
Ilias, F., *et al.* 2020, JINST, 15, 6

Machine Learning in Astronomy: Possibilities and Pitfalls
Proceedings IAU Symposium No. 368, 2025
J. McIver, A. Mahabal & C. Fluke, eds.
doi:10.1017/S1743921323000716

Unsupervised clustering visualisation tool for Gaia DR3

Marco Álvarez[1], Carlos Dafonte[1], Minia Manteiga[2], Daniel Garabato[1], Raúl Santoveña[1] and Lara Pallas[1]

[1]CITIC - Computer Science and IT, University of A Coruña, Spain
email: marco.antonio.agonzalez@udc.es

[2]CITIC - Nautical Sciences and Marine Engineering, University of A Coruña, Spain

Abstract. The Gaia mission DR3 provides accurate data of around two billion stars in the Galaxy, including a classification based on astronomical classes of objects. In this work we present a web visualization tool to analyze one of the products published in the DR3, the Outlier Analysis Self-Organizing Map†.

Keywords. Gaia, DR3, Self-Organizing Maps, Visualization, Big Data

1. Content

Gaia mission is providing accurate data on the astrometric and spectrophotometric properties of around two billion stars in the Galaxy, and several hundred million extra-galactic objects. DPAC international consortium oversees data processing and handles delivering usable scientific information to the community.

Among the products provided in the archive is the classification of the sources into the main astronomical classes of objects and an analysis of the classification outliers. For this purpose, our group has developed a technique based on unsupervised learning, Self-Organizing Maps (SOM), which provides a grouping of these objects into element samples whose common properties are used to provide prototypical objects and statistics of Gaia measurements on them.

SOM provides excellent quality grouping as well as a non-linear dimensionality reduction, by projecting the input data in a determined number of neurons or clusters, usually arranged in a two-dimensional lattice, preserving the topology, see Figure 1. One of the main difficulties of this technique relies on the analysis of the maps, related to the physical and statistical properties of the clustered objects, containing neurons populated with millions of inputs in a Big Data scenario like Gaia. To ease an in-depth study of the content of the map and their neurons, we developed a visualisation tool, GUASOM‡, published with the DR3 of Gaia.

This tool has been developed using Web technologies in a client-server architecture, in which, the frontend has been developed as a Web application using Angular and the backend has been developed as a REST server using Spring.

Concurrent access is expected, involving a wide variety of data in a Big Data scenario that must be delivered in short periods of time, and for that reason, all the data involved must be prepared in advance in simple, sorted, and efficient data structures having only the information requested by the users.

† Data available at `https://cdn.gea.esac.esa.int/Gaia/gdr3/`
‡ GUASOM is available at `https://guasom.citic.udc.es`

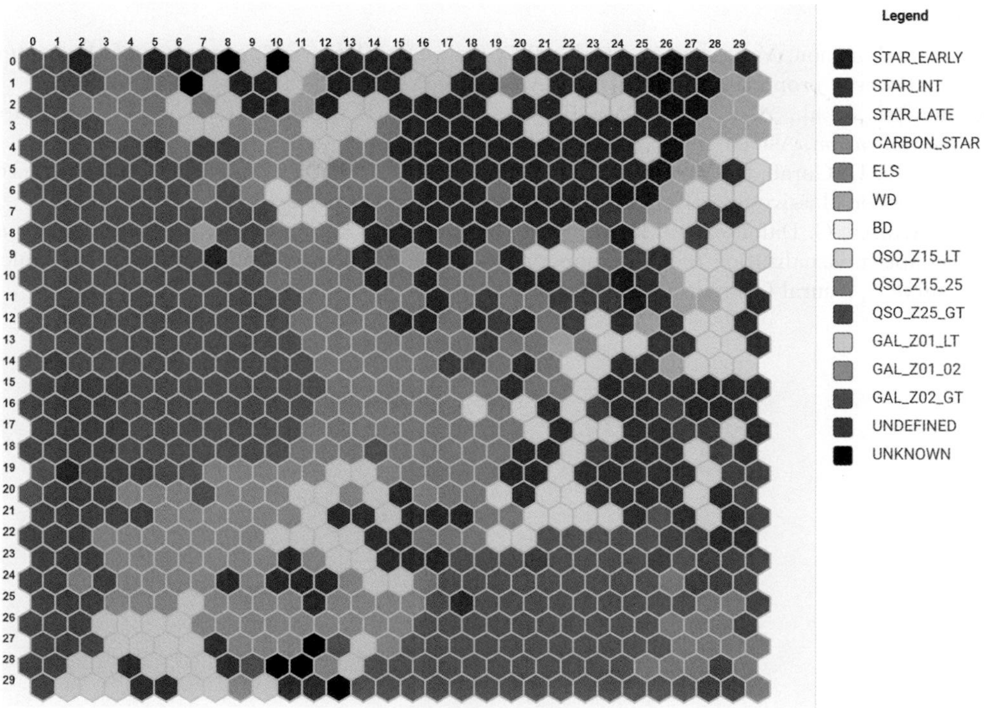

Figure 1. Classification of the objects that belong to each neuron using a reference dataset.

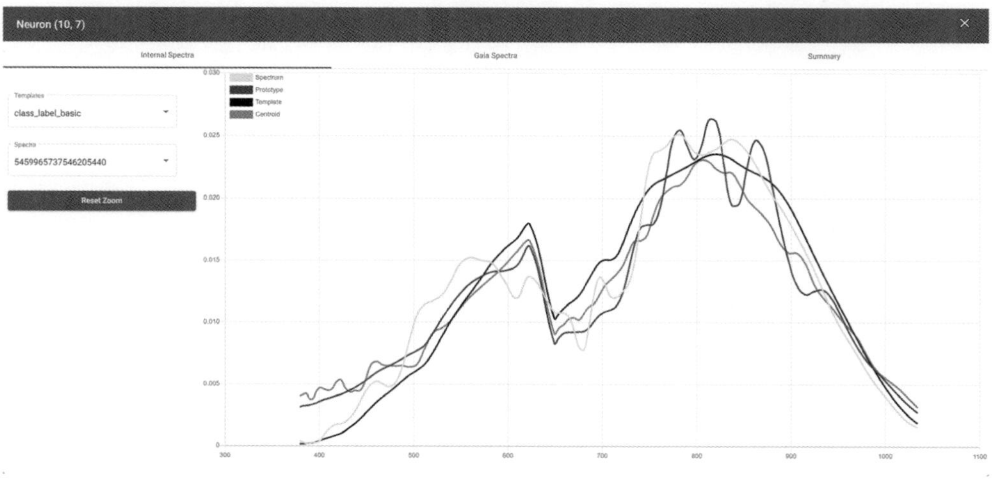

Figure 2. Main characteristics of a neuron, including the prototype, the centroid, the matching template and the spectrum of a particular source that belongs to the neuron.

Moreover, this tool must supply information that allows the scientific personnel to analyse the data behind the scenes, and to do that, it is mandatory to provide some specialized visualizations that highlight the physical content of the data obtained from the Gaia catalogue as well as other well-known catalogues through crossmatch procedures. See Figure 2.

References

Gaia Collaboration, Vallenari, A., et al., 2022, Gaia Data Release 3. Summary of the content and survey properties, Astronomy & Astrophysics

Delchambre, et al., 2022, Gaia Data Release 3: Apsis III - Non-stellar content and source classification, Astronomy & Astrophysics

C. Dafonte, D. Garabato, M. Álvarez and M. Manteiga, 2018, Distributed Fast Self-Organized Maps for Massive Spectrophotometric Data Analysis. Sensors 18 (5), pp. 1419

M. A. Álvarez, C. Dafonte, M. Manteiga, D. Garabato and R. Santoveña, 2021, GUASOM: an adaptive visualization tool forunsupervised clustering in spectrophotometric astronomical surveys, Neural Computing and Applications

Machine Learning in Astronomy: Possibilities and Pitfalls
Proceedings IAU Symposium No. 368, 2025
J. McIver, A. Mahabal & C. Fluke, eds.
doi:10.1017/S1743921322005051

Kinematic Planetary Signature Finder (KPSFinder): convolutional neural network-based tool to search for exoplanets in ALMA data

Jaehan Bae

Department of Astronomy, University of Florida, Gainesville, FL 32611, USA
email: jbae@ufl.edu

Abstract. One of the best and most direct ways to study planet formation processes is to observe young planets while they are forming within their birth protoplanetary disks. As they form, planets tidally interact with their parental disk and produce observable signatures. Recent observations have demonstrated that kinematic planetary signatures (KPS), the perturbed velocity fields of the gas in the protoplanetary disk in the vicinity of the planet, can be observed with the Atacama Large Millimeter/submillimeter Array (ALMA). Here, I introduce a machine learning-based tool KPSFinder (Kinematic Planetary Signature Finder), which aims to find KPS robustly and efficiently.

Keywords. exoplanets, protoplanetary disks, radio interferometry, machine learning, convolutional neural network

---

1. Introduction

Thanks to increasingly powerful observing facilities and techniques, we now routinely find *substructures* in protoplanetary disks, such as annular rings and gaps, and spirals, and vortices (see e.g., review by Andrews 2020). Of particular interest for this project are the perturbed velocity fields of the gas in the protoplanetary disk in the vicinity of the planet, the so-called kinematic planetary signatures (KPS; Pinte et al. 2018, 2022). In the absence of perturbations, the dynamics of a protoplanetary disk is dominated by the Keplerian rotation which manifests as the characteristic butterfly pattern in molecular line observations (Figure 1 left). When there is a planet in the disk, the planet tidally interacts with the disk, launching spiral arms. The spiral arms perturb the disk gas and this perturbation manifests as a discontinuity in the butterfly pattern (Figure 1 middle). Figure 1 right presents a CO channel map of a young star HD 163296, showing a perturbed velocity field that is strikingly similar to the expected kinematic feature arising from planet-disk interactions. Based on forward modeling, it is suggested that an embedded planet with an estimated mass of about 2 $M_{\rm Jup}$ could have driven the KPS-like feature in the disk (Pinte et al. 2018).

Until now, searching for KPS has been relying on visual inspection: one goes through a data cube while looking for features departing from the smooth, Keplerian rotation. Visual inspection has been successful for a few of the most straightforward cases (e.g., Pinte et al. 2018, 2019); however, many other cases remain inconclusive (e.g., Pinte et al. 2020). Furthermore, in order to estimate the mass of the planet associated with the KPS, one needs to run multiple three-dimensional planet-disk interaction simulations, post-process the simulation outputs by running radiative transfer calculations, and make

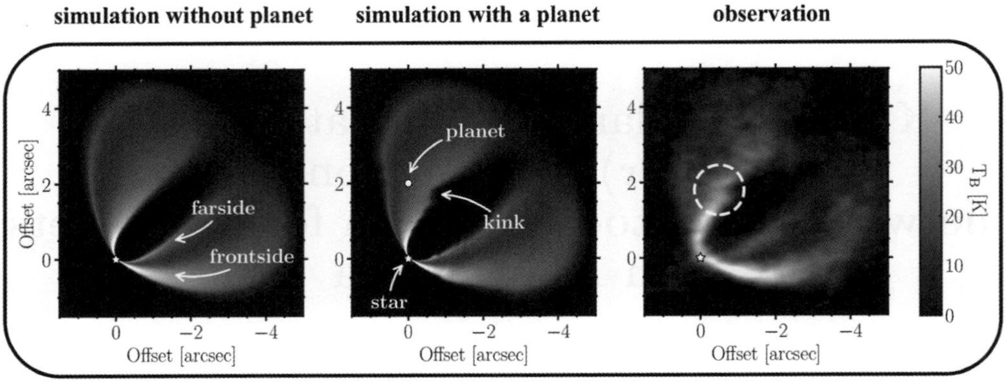

Figure 1. (Left) A simulated CO channel map in the absence of a planet, showing the characteristic butterfly pattern dominated by the Keplerian rotation. The color scale shows the brightness temperature. The star is located at offset = $(0'', 0'')$. (Middle) A simulated CO channel map based on a hydrodynamic simulation with a 2 $M_{\rm Jup}$ planet, at the same velocity channel as in the left panel. The velocity structure of the disk gas in the vicinity of the planet is perturbed, manifesting as a discontinuity, or a 'kink', in the butterfly pattern. (Right) A CO channel map of HD 163296 from ALMA observation (reproduced from Pinte et al. (2018)), showing the characteristic kinematic feature expected from planet-disk interactions (dashed white circle).

a thorough comparison between the simulated data and the observational data, which are very time-consuming even for experts in such tasks.

2. KPSFinder

As high-quality ALMA data reveal fine details of protoplanetary disks, it is crucial to have a method that can more efficiently and robustly detect KPS. Thankfully, the required task of searching for specific features in three-dimensional image cubes has been widely implemented in the field of machine learning, and we can make use of existing open-source machine learning algorithms that are thoroughly tested using a broad range of data sets for various implications. In particular, convolutional neural networks (CNNs; Lecun et al. 1998) are designed to adaptively learn spatial hierarchies of features and are used to find specific features in images.

We implement convolutional neural networks to search for KPS in ALMA line data. The KPSFinder is developed through four main tasks. First, we carry out three-dimensional planet-disk interaction simulations varying the mass of the planet and the disk temperature, the two most important factors in determining the magnitude of KPS, along with the radial location of the planet. Second, using the density and velocity fields obtained from the planet-disk interaction simulations, we carry out radiative transfer calculations to generate synthetic CO data cubes. We vary the inclination of the disk and the azimuthal location of the planet with respect to the disk's position angle. Third, we create training images using the synthetic CO cubes. In order to populate the training images, we apply random rotation to the synthetic CO cubes. We then convolve the cubes with a finite Gaussian beam and add random noise. Finally, we train convolutional neural networks using the training images. The final models tell users whether or not an image cube has KPS. If the image cube has KPS, the models tell users where in the cube the KPS is located and what the estimated planet mass is.

With KPSFinder in hand, users will be able to check whether their data cube contains KPS in a few minutes with a single CPU on a personal computer, without needing to know the details of machine learning algorithms, train/evaluate the algorithm by themselves, or carry out time-consuming forward modeling. This will enable the community to efficiently detect young, forming exoplanets in ALMA data.

Supplementary material

To view supplementary material for this article, please visit https://doi.org/10.1017/S1743921322005051

References

Andrews, S. 2020, ARA&A, 58, 483
Pinte, C. et al. 2018, ApJL, 860, 13
Pinte, C. et al. 2022, arXiv:2203.09528
Pinte, C. et al. 2019, NatAs, 3, 1109
Pinte, C. et al. 2020, ApJL, 890, 9
Lecun, Y. et al. 1998, Proceedings of the IEEE, 86(11):2278–2324

Machine Learning in Astronomy: Possibilities and Pitfalls
Proceedings IAU Symposium No. 368, 2025
J. McIver, A. Mahabal & C. Fluke, eds.
doi:10.1017/S1743921323000571

Predicting Physical Parameters of Cepheid and RR Lyrae variables in an Instant with Machine Learning

A. Bhardwaj[1], E. P. Bellinger[2], S. M. Kanbur[3] and M. Marconi[1]

[1]INAF-Osservatorio Astronomico di Capodimonte, Salita Moiariello 16, 80131, Naples, Italy
email: anupam.bhardwaj@inaf.it

[2]Max Planck Institute for Astrophysics, Karl-Schwarzschild-Straße 1, 85748, Garching, Germany

[3]Department of Physics, State University of New York, Oswego, NY 13126, USA

Abstract. We present a machine learning method to estimate the physical parameters of classical pulsating stars such as RR Lyrae and Cepheid variables based on an automated comparison of their theoretical and observed light curve parameters at multiple wavelengths. We train artificial neural networks (ANNs) on theoretical pulsation models to predict the fundamental parameters (mass, radius, luminosity, and effective temperature) of Cepheid and RR Lyrae stars based on their period and light-curve parameters. The fundamental parameters of these stars can be estimated up to 60 percent more accurately when the light-curve parameters are taken into consideration. This method was applied to the observations of hundreds of Cepheids and thousands of RR Lyrae in the Magellanic Clouds to produce catalogs of estimated masses, radii, luminosities, and other parameters of these stars.

Keywords. stars: variable, Cepheid, RR Lyrae - galaxies: Magellanic Clouds - cosmology: distance scale

1. Introduction

Classical Cepheids and RR Lyrae variable stars are well known distance indicators and are also excellent tracers of young and old age stellar populations, respectively (Subramanian et al. 2017; Beaton et al. 2018; Bhardwaj 2020, 2022). Modern pulsation codes can reproduce observed pulsation periods, light and radial velocity curves, and peak-to-peak amplitude variations for these classical pulsating stars at all wavelengths (Bono, Marconi & Stellingwerf 2000; Marconi et al. 2013, 2015). In the era of large variability surveys, a quantitative comparison of the predicted and observed pulsation properties of Cepheid and RR Lyrae variables provides stringent constraints on their intrinsic evolutionary parameters, and subsequently, provides new challenges for the stellar evolution and pulsation theory (Bhardwaj et al. 2015, 2017; Das et al. 2018). We aim to employ modern automated methods for comparing observed and predicted light curve structure, which is defined by its amplitude and phase parameters, of Cepheid and RR Lyrae variables and generate catalogs of physical parameters of observed variables in the Galaxy and the Magellanic Clouds.

2. Analysis and results

Our theoretical model grid includes 390 models with composition (Y = 0.25 and Z = 0.008) representative of Cepheids in the Large Magellanic Cloud (LMC) computed

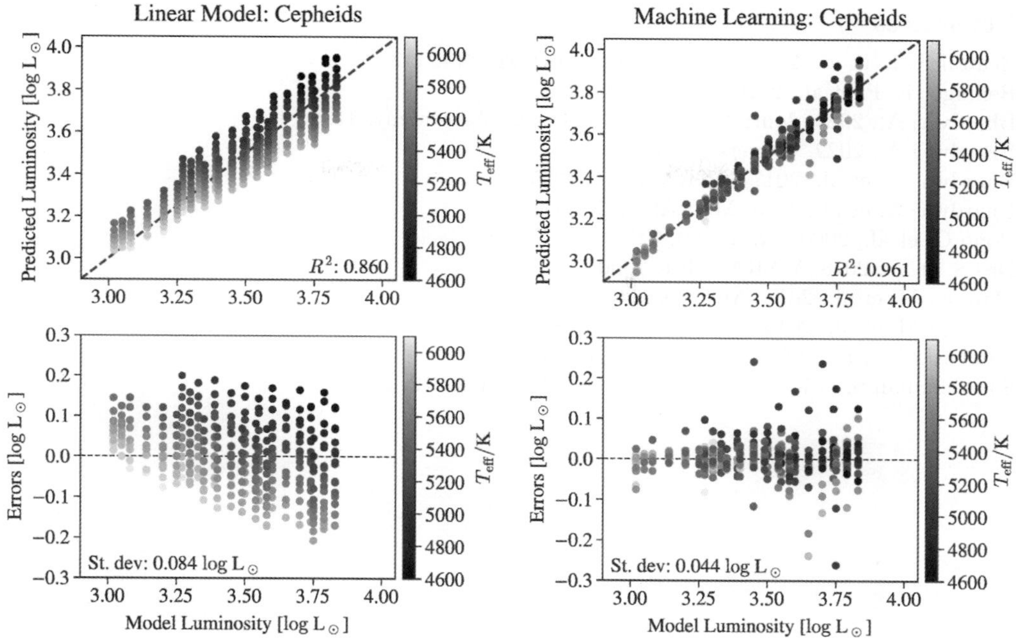

Figure 1. Predicted versus actual luminosities for estimates based on the period (left) and period and light curve parameters (right). The trends in errors (left) vanish when the light curve structure is included (right).

by Marconi et al. (2013), and a total of 270 models representative of RR Lyrae metallicities ranging from Z = 0.0001 to Z = 0.02 and helium abundances ranging from Y = 0.245 to Y = 0.27 (Marconi et al. 2015). The pulsation models provide multi-band theoretical light curves for a broad range of stellar masses, luminosity levels, and chemical composition. The observational $V-$ and $I-$band light curves were taken from the Optical Gravitational Lensing Experiment survey (Soszyński et al. 2018). We use ANNs trained using *scikit-learn* on the grid of theoretical models to predict physical parameters of observed stars (see Bellinger et al. 2020, for more details). The light curve structure importance was independently verified using random forests.

We evaluate how well the ANN can predict physical parameters based on the theoretical models and using: 1) period (linear model); 2) period and light curve structure (machine learning method). The two-fold cross validation method was used for model assessment and a quantification of improvement was provided using the standard deviation of the errors and the coefficient of variation (R^2) - a higher R^2 is better. In Figure 1, a significant increase in R^2 and a large reduction in the standard deviation of residuals is clearly seen when using the machine learning method. Similar results were also obtained for masses, radii, temperatures, absolute magnitudes, colors, and Wesenheit magnitudes (Bellinger et al. 2020).

3. Conclusions

While the period-mean density relation suggests that the observable period is the most important quantity in constraining the global stellar parameters of classical pulsating stars, we showed that the light curve structure plays a statistically significant part in determining these parameters. Once a smooth grid of pulsation models covering the entire parameter space is available, the catalogs of physical parameters can be instantly generated for all observed Cepheid and RR Lyrae stars with unprecedented precision.

References

Beaton R. L. et al., 2018, Space Science Reviews, 214, 113
Bellinger E. P. et al., 2020, MNRAS, 491, 4752
Bhardwaj A., 2020, Journal of Astrophysics and Astronomy, 41, 23
Bhardwaj A., 2022, Universe, 8, 122
Bhardwaj A. et al., 2017, MNRAS, 466, 2805
Bhardwaj A. et al., 2015, MNRAS, 447, 3342
Bono G. et al., 2000, A&A, 360, 245
Das S. et al., 2018, MNRAS, 481, 2000
Marconi M. et al., 2015, ApJ, 808, 50
Marconi M. et al., 2013, MNRAS, 428, 2185
Soszyński I. et al., 2018, Acta Astron., 68, 89
Subramanian S. et al., 2017, Space Science Reviews, 212, 1817

Machine Learning in Astronomy: Possibilities and Pitfalls
Proceedings IAU Symposium No. 368, 2025
J. McIver, A. Mahabal & C. Fluke, eds.
doi:10.1017/S1743921322003532

Bayesian deconvolution of a rotating spectral line profile to a non-rotating one

M. Curé[1], **P. Escarate**[2], **L. Celedon**[1], **J. Cavieres**[3], **E. Olivares**[2], **I. Araya**[4], **C. Arcos**[1], **R. Pezoa**[5], **G. Farias**[2] **and N. Machuca**[1]

[1]Instituto de Física y Astronomía, Universidad de Valparaíso
email: michel.cure@uv.cl.

[2]Escuela de Ingeniería Eléctrica, Pontificia Universidad Católica de Valparaíso

[3]Instituto de Estadística, Universidad de Valparaíso

[4]Vicerrectoría de Investigación, Universidad Mayor

[5]Escuela de Ingeniería Informática, Universidad de Valparaíso

Abstract. In this work we present a Bayesian approach to obtain the non-rotating stellar spectra from an observed spectrum of a rotating star. This is our first attempt to solve an inverse problem expressed in terms of a Fredholm integral. Our preliminary results with synthetic spectra are promising. More studies are required to compare our Bayesian approach with the standard method for real spectra.

Keywords. stars: rotation, methods: numerical, methods: statistical, techniques: spectroscopic

---

1. Introduction

Nowadays, one of the standard procedures to determine stellar and wind parameter of massive stars, is to compare the observed spectral lines with synthetic ones which are calculated using state-of-the-art Non Local Thermal Equilibrium (NLTE) radiative transport codes, with hydrodynamic velocity and density profiles used as input. We developed the `ISOSCELES` grid (Araya et al. in prep) of 1/2 million of models for OBA types by using the radiation driven wind theory and a NLTE radiative transport code to obtain the synthetic lines. All these models are calculated for a non-rotating star. Thus, to compare our observed line with `ISOSCELES`, we need to rotationally convolve the synthetic lines, a very time consuming task that must be repeated whenever another star (with a different rotational speed) is analyzed. In this work, we propose a methodology to deconvolve the observed line to a line of a non-rotating star. Therefore, we can use different machine learning techniques applied only to the 'original' (non-rotating) grid, sparing significant time resources.

2. Numerical Method

This is a very difficult inverse problem described by Carroll (1993) and our proposed deconvolution method is to transform this to the optimization of a direct problem. We propose that the non-rotating line can be described by a superposition of Gaussian functions. Then, we convolve these Gaussians (with the measured rotational speed) to obtain the rotating line and compare it directly with the observed line. We implemented this procedure using the Hamiltonian Monte Carlo code "Stan" (Stan Development Team 2022) to solve this problem. The encouraging results for emission, absorption and P-Cyg

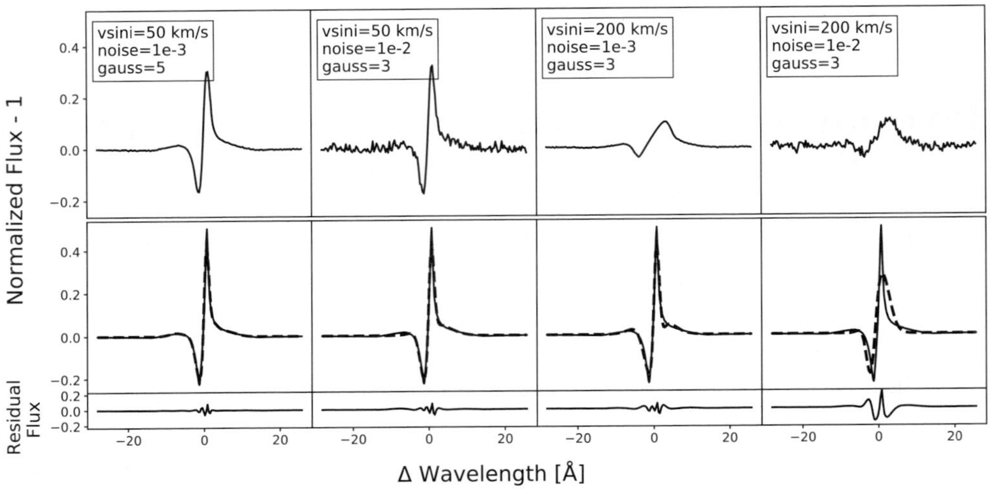

Figure 1. Example of a P-Cyg Hα synthetic line deconvolution. The synthetic line was first convoluted using different values of $v \sin i$ and noise level (top row) before being deconvolved using a 3-Gaussian profile (5 for the first model). Recovered (dashed line) and original non-rotating synthetic (black solid line) profiles are shown in middle row, while the residuals are displayed in the bottom row.

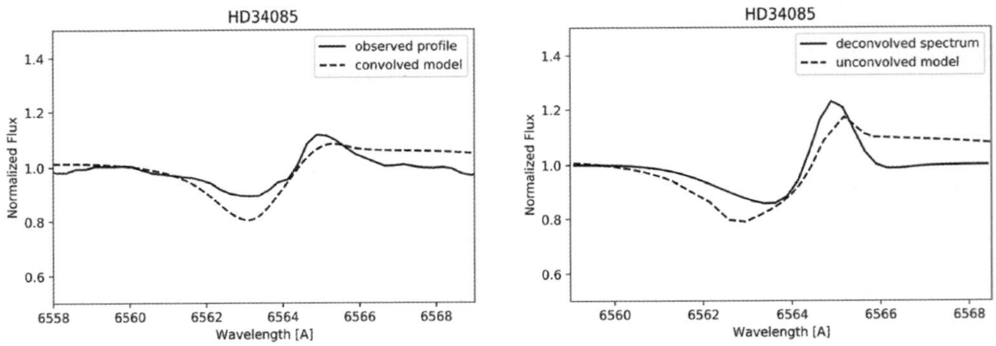

Figure 2. Left panel: Standard procedure for the star HD34085, observed Hα profile (black solid line) and the best rotational convolved model (dashed line) from `ISOSCELES` database; $T_{\rm eff} = 17500$K, $\log g = 2.25$, $\dot{M} = 5 \times 10^{-7}$ $M_\odot$/yr and $v_\infty = 446$ km/s. Right panel: Same star, but now the deconvolved spectrum (black solid line) and the best unconvolved model (dashed line) from `ISOSCELES` ; $T_{\rm eff} = 17500$K, $\log g = 2.25$, $\dot{M} = 8 \times 10^{-7}$ $M_\odot$/yr and $v_\infty = 576$ km/s.

lines present ubiquitously in the spectra of massive stars spectra are shown, also some results for stellar and wind parameters of a sample of massive stars. For future work, we can apply convolutional neural networks to learn the grid of models and rapidly find stellar and wind parameters.

3. Discussion about preliminary results

Our first results are very promising but more comparisons between the new Bayesian methodology with the ones calculated using the standard procedure are needed. In addition, we need to improve our Stan code to be more efficient, in its current state it can only works with low resolution spectra within an acceptable time. We also developed a frequentist code, and we will compare the performances of both approaches. Finally, we want to include the microturbulence velocity in the rotational profile, as well as the simultaneous deconvolution of several lines.

Acknowledgments
Authors are grateful to project FONDECYT 1190485. CA, RP, IA and GF thanks to FONDECYT projects: 11190945, 3190740 (postdoc), 11190147 and 1191188, respectively. This project has also received funding from the European Union's Framework Programme for Research and Innovation Horizon 2020 (2014–2020) under the Marie Skłodowska-Curie grant Agreement No 823734.

References

Carroll, J. A. 1933, MNRAS, 93, 478.
Stan Modeling Language Users Guide and Reference Manual, 2.30 (2022). https://mc-stan.org

Machine Learning in Astronomy: Possibilities and Pitfalls
Proceedings IAU Symposium No. 368, 2025
J. McIver, A. Mahabal & C. Fluke, eds.
doi:10.1017/S1743921322003635

A short study on the representation of Gravitational Waves Data for Convolutional Neural Network

M. Grespan

National Centre for Nuclear Research, Astrophysics division
ul. Pasteura 7, 02-093 Warszawa
email: margherita.grespan@ncbj.gov.pl

Abstract. Binary Neutron star mergers (BNS) are of particular interest for the possibility of multi-messenger astronomy. We investigate if BNS gravitational waves trigger detection should be preferably conducted in a representation instead of another, we inspect time strains, Fourier transform, and spectrograms. We recreate the data stream of the LIGO detector, which could contain noise only or signals hidden in the noise. We train a binary classifier able to distinguish if a merger is present. We create two distinguished datasets, one with the LIGO detector instrumental noise and one whitened with respect to the detector PSD. We obtain the highest accuracy for whitened spectrograms - 93% correct classifications of an external validation dataset. Fourier transforms show similar behavior in both colored and whitened datasets with an accuracy of ∼77%. Whitened time strains have a substantial increase in accuracy from the colored to the whitened dataset, respectively 50% and 73%.

Keywords. gravitational waves, methods: statistical, methods: data analysis

1. Introduction

The LIGO, Virgo, KAGRA network has reached enough sensibility to expect one detection per day (Abbott 2020). However, the current available methods for GW detection and data analysis are computationally expensive, not permitting a fast response to potential triggers. Time constrains have an important role in multimessenger astronomy, if the detection is not fast enough it is possible to miss a multi-messenger observation opportunity. Neural Networks (NN) are an interesting candidate for this job for their computational efficiency once trained. In the last years machine learning algorithms have been used as a tool in GW astrophysics with a variety of applications, from GW detection to parameter estimation, denoising of GW signals, glitches and noise transients classification. Most of these studies involve the use of convolutional neural networks (CNN), but with different data representations. In the case of signal detection, 9 publications use time-domain data (Schafer 2020; George 2018, 2017; Gabbard 2018; Krastev 2020; Luo 2019; Li 2020; Fan 2019) and 3 (Wei & Wu 2017; Morales 2021; Marulanda 2020) use the spectrogram representation. We investigate if there is a GW representation, for a simulated Binary Neutron Star (BNS) waveform injected in the LIGO noise, with a better performance for a simple CNN. The choice of using BNS waveforms is based on the knowledge that, if at least a neutron star is present in a binary system, there is the possibility to observe electromagnetic counterparts.

Table 1. Results for the the two datasets.

Signal	Representation	Accuracy
Colored	Time	0.5
	Spectrogram	0.91
	FFT	0.77
Whitened	Time	0.73
	Spectrogram	0.93
	FFT	0.78

2. Methods

BNS mergers reach higher frequencies and spend more time in the sensitive band of the detector (Schafer 2020), data need to be sampled at a higher rate giving a high amount of data to analyze. We investigate the simplest case of establishing whether a signal is present or not in a 10 s strain of data. We label with 1 data containing a merging (signal + noise) and 0 the noise only. We simulate mergers with IMRPhenomD_NRTidal approximant, with a sample rate of 1024 Hz, masses ranging from 1 to 2 $M_\odot$ in steps of 0.1 $M_\odot$, a lower frequency cut-off of 40 Hz, distances from the observer from 10 Mpc up to 300 Mpc with 2 Mpc step, zero individual spins, no orbital eccentricities, and zero inclination. Mass and distance combinations are picked randomly. We crop 10 s time strains containing the signal to noise ratio peak (pSNR). Each waveform is positioned randomly in the noise, after the peak, we leave a random amount of seconds that goes from 0.25 to 0.75 and the remnant part of the signal is cut accordingly to have a 10 s strain. The dataset contains only pSNR in a 5 to 50 interval. With this procedure, we create two datasets, one with the LIGO detector noise and one cleaned accordingly the instrument PSD. We test the network with a minimum of 5 layers to a maximum of 10, with, different dropouts, pooling, and kernel sizes. Pooling layers have size 2 and kernel size 4. The two datasets contain $\sim$10.000 strains, equally divided into the two classes, 20% of it is used as a test set, and the remnant is used for training. We furthermore generate an external validation dataset with $\sim$2000 elements equally split in noise and signal+noise.

3. Results and Conclusions

In Tab. 1 are shown the final results for the best models. The network is not able to distinguish between signal and noise in the time representation of the colored dataset. However, the same architecture in the whitened dataset reaches 70% accuracy. The overall best model is provided by the whitened spectrogram dataset. Spectrograms have the best accuracy, but in literature, only three works use this representation against the nine time series (see the introduction for references). This difference could be due to the spectrogram loss of information with respect to the time data, for parameter estimation, this could make a difference. However, we think that for trigger warnings spectrograms are a good choice, they require less memory and provide better results.

Supplementary material

To view supplementary material for this article, please visit https://doi.org/10.1017/S1743921322003635

References

Abbott et al.2020, *Springer Science and Business Media LLC*, 23
George, D. & Shen H. & Huerta, E. A.,2017, *Phys. Rev. D*, 97

Gabbard H. & Williams M. & Hayes F. and Messenger C.,2018, *Phys. Rev. Lett.*, 120
Krastev P. G., 2020 *Phys. Rev. Lett B*, 803
Morales M. D. & Antelis J. M. & Moreno C. & Nesterov A. I., 2021*Sensors*, 21, 3174.
Marulanda J. P. & Santa C. & Romano A. E., 2020*Phys. Rev. Lett. B*, 810, 3174.
George, D. & Huerta, E. A.2018, *Phys. Rev. Lett B* 778
Wei J. & Wu X., 2017, *MNRAS*, 472
Morales, M. D. & Antelis J. M. & Moreno C. & Nesterov A. I., 2021,*Sensors*, 21
Schafer M. B. & Ohme F. & Nitz A. H., 2020,*Phys. Rev. D*, 102, 6
H. Luo & W. Lin & Z. Chen & Q. Huang, 2019,*Frontiers of Physics*, 15, 1
Li X. & Yu W. & Fan X. & Jogesh Babu G., 20120,*Frontiers of Physics*, 15, 5
Fan X. & J. Li & X. & Y. Zhong & Junwei Cao, 2019,*Science China Physics,Mechanics & Astronomy*, 15, 5

Machine Learning in Astronomy: Possibilities and Pitfalls
Proceedings IAU Symposium No. 368, 2025
J. McIver, A. Mahabal & C. Fluke, eds.
doi:10.1017/S1743921323000765

Search for Microlensing Signature in Gravitational Waves from Binary Black Hole Events

Kyungmin Kim

Department of Physics, Ewha Womans University, Seoul 03760, South Korea
email: kmkim82@gmail.com

Abstract. In a recent search (Kim et al. 2022), we looked for microlensing signature in gravitational waves from spectrograms of the binary black hole events in the first and second gravitational-wave transient catalogs. For the search, we have implemented a deep learning-based method (Kim et al. 2021) and figured out that one event, GW190707_093326, out of forty-six events, is classified into the lensed class. However, upon estimating the p-value of this event, we observed that the uncertainty of the p-value still includes the possibility of the event being unlensed. Therefore, we concluded that no significant evidence of beating patterns from the evaluated binary black hole events has found from the search. For a consequence study, we discuss the distinguishability between microlensed GWs and the signal from precessing black hole binaries.

Keywords. Gravitational waves, Gravitational microlensing

1. Previous Searches

Similar to light, gravitational waves (GWs) can be lensed when they propagate near massive astrophysical objects. To date, many searches have tried to look for the signature of the gravitational lensing of GWs (GW lensing, hereafter) (Kim et al. 2022, and references therein). However, there has been no widely accepted detection thus far.

Alternatively, in a recent search (Kim et al. 2022), we have revisited the forty-six BBH events analysed in Hannuksela et al. (2019); Abbott, R. et al. (2021) and evaluated them with the deep learning-based classification method built in an earlier work (Kim et al. 2021): We implement one of the state-of-the-art deep learning models, VGG-19 (Simonyan & Zisserman 2014) by setting the batch size to be 128, the maximum training epoch to be 10 with adopting the Adam optimizer (Kingma & Ba 2014) and the cross-entropy loss for the training. We newly introducing ensemble learning to mitigate possible biased prediction from a single learner. On top of that, we have built a new hierarchical classification scheme for the evaluation of BBH signals. One can refer Kim et al. (2021, 2022) and the attached supplementary material for more details of utilizing the deep learning model.

From the search, we figured out only one event, GW190707_093326, out of forty-six events is primarily classified as lensed. To verify the result, we estimated the p-value for the median probability $\bar{r} = 0.984^{+0.012}_{-0.342}$ and observed that the uncertainty of the estimated p-value, ranging $0 \lesssim p \lesssim 0.1$, is still containing $p \geq 0.05$ where the unlensed hypothesis is true. Therefore, we have concluded that the search has found no significant evidence of beating patterns from the evaluated BBH events.

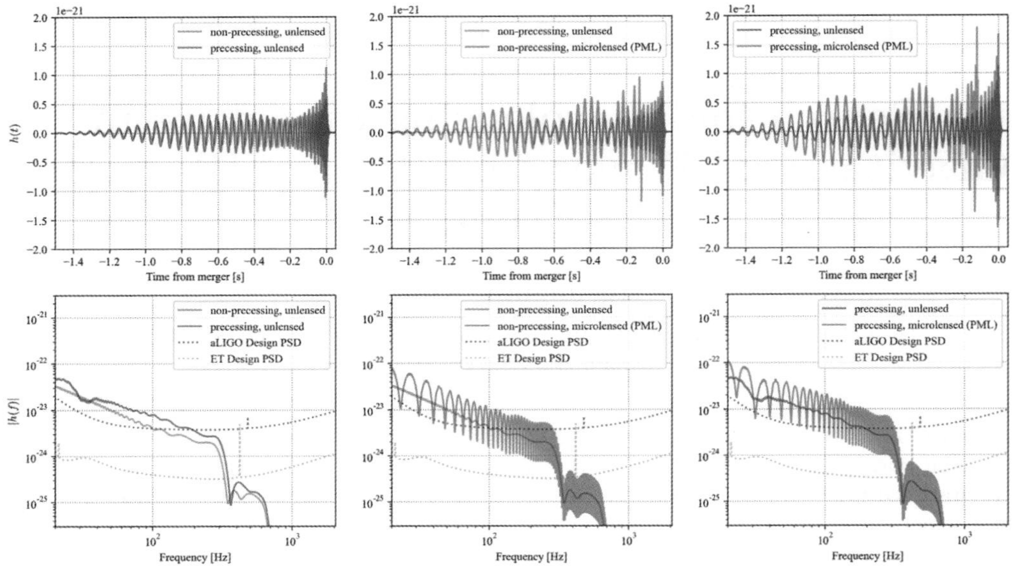

Figure 1. Example waveforms showing how different a microlensed GW signal against the signal of a precessing BBH merger. We assume the same component mass ($m_1 = m_2 = 30M_\odot$) and the same distance to the source (500Mpc) for both cases. For the generation of the signal from precessing BBH merger, we set arbitrary spin vectors for each component black hole. For microlensed signals, we adopt the point-mass lens model in the geometrical optics limit with the source position parameter $y = 0.3$ and the redshifted lens mass $M_{lz} = 10^4 M_\odot$. (see Takahashi & Nakamura (2003) for details of lensing configuration and formulation.)

2. Can we distinguish microlensed GWs from the signal of precessing BBH mergers?

In the two previous works, we have supposed the microlensing of non-precessing GWs for convenience. However, it is known that precessing BBH systems can introduce modulation in the waveform which looks quite similar to the beating pattern shown in the waveform of microlensed GWs. In Fig. 1, we present example waveforms corresponding to the cases: The left panels show how the waveforms from non-precessing (green) and precessing (blue) look different in the time domain (top) and frequency domain (bottom). In the middle and right panels, we provide waveforms of microlensed signals of non-precessing and precessing BBH mergers, respectively. From the comparison between unlensed precessing waveform and lensed non-precessing waveform in the time domain (the top panels of the left and middle panels, respectively), we can observe the modulations shown in the inspiral phase are rather different. In the frequency domain, the modulation from the precession appears as a smooth sinusoidal oscillation while the beating pattern by the microlensing produces arch-shaped patterns. Moreover, in the time-domain waveform, we can see there are two peaks between $[-0.2, 0]$ seconds for the lensed signal. However, when we compare the middle and right panels, the GW microlensing from the same lensing configuration introduces very similar modulations which hinders to identify whether it came from non-precessing or precessing binaries.

From this preliminary qualitative study, it looks feasible to certainly distinguish the microlensing-induced beating patterns from the precession-induced modulation. However, it will be quite challenging to distinguish whether the signal is from non-pressing or precessing binaries when both cases are microlensed. We expect exploiting a machine learning method such as the distance metric learning will be a possible method for an exploration of the distinguishability even in noisy data in terms of the similarity between the target of interest.

Supplementary material

To view supplementary material for this article, please visit https://doi.org/10.1017/S1743921323000765

References

Abbott, R., Abbott, T. D., Abraham, S. et al. 2021, Astrophys. J., 923, 14, doi: 10.3847/1538-4357/ac23db

Hannuksela, O., Haris, K., Ng, K., et al. 2019, Astrophys. J. Lett., 874, L2, doi: 10.3847/2041-8213/ab0c0f

Kim, K., Lee, J., Yuen, R. S. H., Hannuksela, O. A., & Li, T. G. F. 2021, Astrophys. J., 915, 119, doi: 10.3847/1538-4357/ac0143

Kim, K., Lee, J., Hannuksela, O. A., & Li, T. G. F. 2022. Astrophys. J., 938, 157, doi: 10.3847/1538-4357/ac92f3

Kingma, D., & Ba, J. 2014, International Conference on Learning Representations

Simonyan, K., & Zisserman, A. 2014, ArXiv e-prints. https://arxiv.org/abs/1409.1556

Takahashi, R., & Nakamura, T. 2003, Astrophys. J., 595, 1039. http://stacks.iop.org/0004-637X/595/i=2/a=1039

Machine Learning in Astronomy: Possibilities and Pitfalls
Proceedings IAU Symposium No. 368, 2025
J. McIver, A. Mahabal & C. Fluke, eds.
doi:10.1017/S1743921322003659

Deep learning and numerical simulations to infer the evolution of MaNGA galaxies

Regina Sarmiento[1,2], **Johan H. Knapen**[1,2], **Marc Huertas-Company**[1,2,6], **Annalisa Pillepich**[3], **Sebastián F. Sánchez**[4], **Héctor Ibarra-Medel**[5] **and Eduardo Lacerda**[4]

[1]Instituto de Astrofísica de Canarias, E-38205 La Laguna, Tenerife, Spain
email: johan.knapen@iac.es

[2]Departamento de Astrofísica, Universidad de La Laguna, E-38205 La Laguna, Tenerife, Spain

[3]Max-Planck-Institut fur Astronomie, Konigstuhl 17, D-69117 Heidelberg, Germany

[4]Instituto de Astronomía, Universidad Nacional Autónoma de México, AP 70-264, CDMX 04510, Mexico

[5]Instituto de Astronomía y Ciencias Planetarias, Universidad de Atacama, Chile

[6]Observatoire de Paris, LERMA, PSL University, 61 avenue de l'Observatoire, F-75014 Paris, France

Abstract. As surveys grow, the challenge is how to explore and interpret the increasing quantity of data. Integral field spectroscopic (IFS) galaxy surveys are a good example of how data have grown in complexity and in volume. In order to find complex relations between the spatially resolved structures of galaxies and their accretion histories, we combine IFS high-dimensionality data, deep learning and numerical simulations to infer the evolutionary paths of galaxies. In this work we generate 10,000 simulated galaxies from TNG50 hydro-cosmological simulation to compare with the 10,000 galaxies observed in MaNGA, thus generating a mock MaNGA sample. We then analyse how the simulated galaxies reproduce the properties of MaNGA galaxies and study how the evolutionary paths of the mock galaxies relate to their observable properties. We finish by outlining our next steps which include using contrastive learning.

Keywords. Galaxies: general, Surveys, Methods: statistical

1. Generating a mock MaNGA

Mapping Nearby Galaxies at APO (MaNGA, Bundy et al. 2015) is currently the largest integral field spectroscopic survey with data for 10,000 nearby galaxies of all types. To infer galaxy evolution from the TNG50 cosmological simulation (Pillepich et al. 2019, Nelson et al. 2019), we emulate the MaNGA survey.

We propose an algorithm to select a simulated pair for each MaNGA galaxy based on their redshift, mass and size. The selected TNG50 galaxies are then processed following Ibarra-Medel et al. 2019 to obtain mock datacubes. Subsequently, these cubes are analysed with pyPipe3D (Sánchez et al. 2016a, Sánchez et al. 2016b, Lacerda et al. 2022) to obtain post-processed maps of age, metallicity and kinematics comparable to those derived from the observed MaNGA galaxies.

2. TNG50 matched sample and post-processed mock cubes

After matching the samples, the distribution of the redshifts, stellar masses and effective radii of the MaNGA galaxies coincide well with those of the matched sample of TNG50 counterparts.

When comparing observed and mock galaxies, we see that the instrumental and pipeline effects produce a realistic mock. Physical properties including stellar mass and its distribution, age, metallicity, line-of-sight velocity and velocity dispersion are properly recovered after the mock generation.

In the distribution across the mass-radius plane, the main trends in the mean values of the age, metallicity and dispersion maps are replicated in the matched TNG50 sample, although MaNGA galaxies are younger and have lower velocity dispersions in the low- to intermediate-mass and large-radius region of the plane. Older galaxies lie in the lower side of the plane, while more metallic galaxies tend to be more massive. As expected, higher velocity dispersion values are found in the more massive end.

3. Next Steps

Future work includes analysing how well TNG50 reproduces the structures in the resolved maps of the observed galaxies. As the history of each simulated galaxy is known, we are able to trace the processes that cause its observational features to infer how observed galaxies evolve. To do this, removing the observational biases and reducing the dimensionality of the data are fundamental. A promising avenue to perform this comparison is an unsupervised Deep Learning algorithm called contrastive learning. Contrastive learning is especially effective with noisy or biased data and produces a representation space that organizes the data by their similarity, as we showed in Sarmiento et al. 2021. There, we used a novel self-supervised machine-learning method to visualize the multidimensional information on stellar population and kinematics in the MaNGA survey in a 2D plane. The framework we used is insensitive to nonphysical properties such as the size of the integral field unit and was therefore able to order galaxies according to their resolved physical properties. Using the extracted representations, we could study how galaxies distribute based on their resolved and global physical properties. We showed that even when exclusively using information about the internal structure, galaxies naturally cluster into two well-known categories: rotating main-sequence disks and massive slow rotators. This was found from a purely data-driven perspective, but confirmed distinct assembly channels.

Because the data complexity is much reduced in this approach, it is also very well suited to compare the simulated and observed galaxies. It can be used to study merging and interacting galaxies, tracing significant events that shape the physical properties of the galaxies.

4. Summary

We built a catalogue comprising 10,000 TNG50 galaxies that match the MaNGA sample. The process to obtain MaNGA-like mocks was implemented on galaxies in this catalogue with positive results. The averaged properties (age, metallicity and velocity dispersion) follow the same trends observed in the MaNGA sample. The complete matched TNG50 sample comprising 10,000 mocks will be soon publicly available. Future work will include the use of contrastive learning to analyse the distribution of common physical structures in the representation space and trace different stages of galaxy evolution from a combination of simulated and observed galaxies.

5. Acknowledgements

This project has received funding from the European UnionÕs Horizon 2020 research and innovation programme under the Marie Sklodowska-Curie grant agreement No 893673, from the State Research Agency (AEI-MCINN) of the Spanish Ministry of Science and Innovation under the grant "The structure and evolution of galaxies and their central regions" with reference PID2019-105602GB-I00/10.13039/501100011033 and under the grant "Thick discs, relics of the infancy of galaxies" with reference PID2020-113213GA-I00, and from IAC project P/300724, financed by the Ministry of Science and Innovation, through the State Budget and by the Canary Islands Department of Economy, Knowledge and Employment, through the Regional Budget of the Autonomous Community.Ê

References

Bundy, K., et al. 2015, ApJ, 798, 7
Ibarra-Medel, H. J., et al. 2019, MNRAS, 483, 4525
Lacerda, E. A. D., et al. 2022, arXiv:Astrophysics - Astrophysics of Galaxies, 97, 101895
Nelson, D., et al. 2019, MNRAS, 490, 3234–3261
Pillepich, A., et al. 2019, MNRAS, 490, 3196–3233
Sánchez, S., et al. 2016a, RMAA, 52, 21–53
Sánchez, S., et al. 2016b, RMAA, 52, 171–220
Sarmiento, R., et al. 2021, ApJ, 921, 177

Machine Learning in Astronomy: Possibilities and Pitfalls
Proceedings IAU Symposium No. 368, 2025
J. McIver, A. Mahabal & C. Fluke, eds.
doi:10.1017/S1743921322003520

Data pre-extraction for better classification of galaxy mergers

W. J. Pearson, L. E. Suelves, NEP Team and GAMA Team

National Centre for Nuclear Research, Pasteura 7, 02-093 Warszawa, Poland
email: william.pearson@ncbj.gov.pl

Abstract. Galaxy mergers are central to our current understanding of the universe. These events are known to change the morphologies of the merging galaxies, accompanied by changes to physical processes Galaxy mergers are rare events and to better study them we need larger, statistical samples.

Greater use of machine learning and artificial intelligence has seen galaxy mergers become a target for these techniques. Recently, application of these techniques has gone from proof of concept to using machine learning galaxy merger catalogues for science.

In this proceedings, we will examine how extraction of morphological parameters by traditional methods can be used to aid neural networks in classification of galaxy mergers. We will see how current networks may not be extracting all the information from imaging data. We will also discuss how samples of rare objects can be contaminated and the impact this has on our science. This **contribution** is based on Pearson *et al.* (2022).

Keywords. Galaxies: interactions – Galaxies: evolution – Methods: data analysis – Galaxies: statistics

1. Introduction

Galaxy mergers underpin our current understanding of how galaxies grow and evolve. To have a better understanding of what happens during a galaxy merger, we need statistical samples. One such way to create these samples is with deep learning. In this proceedings, we will explore how deep learning can be used to identify galaxy mergers in Hyper Suprime-Cam (HSC) Subaru Strategic Program (SSP) and North Ecliptic Pole (NEP) imaging.

2. Merger Identification

We aim to identify galaxy mergers within the HSC coverage of the NEP. To do this, we will use deep learning using a training set of HSC-SSP images of merging and non-merging galaxies as identified in the Galaxy And Mass Assembly- Kilo Degree Survey Galaxy Zoo project (Holwerda *et al.* 2019, Kelvin *et al.* in prep) with an asymmetry-smoothness cut to remove the worst misclassification from Galaxy Zoo. This provides 1683 merging systems at $z < 0.15$ with imaging that is comparable to the NEP region. We train a convolutional neural network (CNN) on the images, a fully connected network (FCN) on the morphological parameters and a combined network joining the CNN and FCN together.

The combined FCN and CNN performed better than either the CNN or FCN alone. Thus, there is information lost by only using the images that is recovered when including the extracted morphologies. We trained a similar network for $0.15 < z < 0.30$.

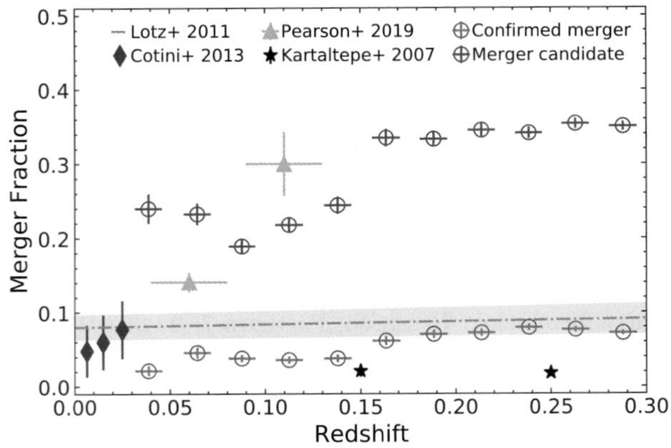

Figure 1. Merger fraction from the neural networks (red circles) and visual classification (blue circles) as a function of redshift. Merger fractions from Cotini *et al.* (2013, purple diamonds), Pearson *et al.* (2019, green triangles), Kartaltepe *et al.* (2007, brown stars) and Lotz *et al.* (2011, dot-dashed purple line) are also shown.

3. Merger misclassification

While the performance of the networks is admirable, with over 85% accuracy, the classifications are imperfect. Our network was applied to 34 264 galaxies in NEP, identifying 10 195 galaxy merger candidates. However, visual inspection of these galaxies found only 2109 of these candidates to be actual mergers. The misclassification is a combination of an imperfect network and galaxy mergers being a minority class. A small fraction of **misclassifications** of the majority class can result in a large fraction of the identified minority class being incorrect.

Figure 1 shows the implications of misclassification. The red circles show the merger fraction of merger candidates, much higher than the observations in other works but **in line** with CNN identified mergers of Pearson *et al.* (2019). The visually confirmed sample is in line with the expected results.

4. Summary

- Using both images and morphologies in neural networks improves classification.
- Even good networks for identifying rare classes are highly contaminated.
- This high contamination has implications for upcoming large area survey.

Acknowledgements

W.J.P. has been supported by the Polish National Science Center project UMO-2020/37/B/ST9/00466 and by the Foundation for Polish Science (FNP).

This paper makes use of software developed for the Large Synoptic Survey Telescope. We thank the LSST Project for making their code available as free software at `http://dm.lsst.org`

This paper is based [in part] on data collected at the Subaru Telescope and retrieved from the HSC data archive system, which is operated by Subaru Telescope and Astronomy Data Center (ADC) at National Astronomical Observatory of Japan. Data analysis was in part carried out with the cooperation of Center for Computational Astrophysics (CfCA), National Astronomical Observatory of Japan.

Based on observations made with ESO Telescopes at the La Silla Paranal Observatory under programme ID 177.A-3016.

References

Cotini, S., Ripamonti, E., Caccianiga, A., *et al.* 2013, *MNRAS*, 431, 2661
Holwerda, B. W., Kelvin, L., Baldry, I., *et al.* 2019, *AJ*, 158, 103
Kartaltepe, J. S., Sanders, D. B., Scoville, N. Z., *et al.* 2007, *ApJS*, 172, 320
Lotz, J. M., Jonsson, P., Cox, T. J., *et al.* 2011, *ApJ*, 742, 103
Pearson, W. J., Wang, L., Alpaslan, M., *et al.* 2019, *A&A*, 631, 51
Pearson, W. J., Suelves L. E., Ho S .C.-C., *et al.* 2022, *A&A*, 661, 52

Machine Learning in Astronomy: Possibilities and Pitfalls
Proceedings IAU Symposium No. 368, 2025
J. McIver, A. Mahabal & C. Fluke, eds.
doi:10.1017/S1743921323001254

Stellar spectra classification and clustering using deep learning

Tomasz Różański

Astronomical Institute, University of Wrocław, Kopernika 11, 51-622 Wrocław, Poland
email: tomasz.rozanski@uwr.edu.pl

Abstract. Most of our knowledge about the Universe comes from the careful analysis of light that reaches us. Spectroscopy, which is the most detailed method of spectrum analysis, when applied to stars provides information on the parameters of their atmospheres, including effective temperature, acceleration, velocity fields, and their chemical composition. Stellar classification brought forth the understanding of what physical parameters are critical in shaping stellar atmospheres. It is a key element that has linked efforts related to numerical modelling of atmospheres with observations. We present preliminary results on the method of stellar spectra classification based on large-scale unsupervised pre-training. The applied deep neural network of the auto-encoder type, thanks to the use of differentiable elements of physical modelling in the decoder, allows to work with medium to high-resolution spectra, is insensitive to normalization errors, and different radial and rotational velocity, and operates in a wide range of signal-to-noise ratio.

Keywords. stellar spectroscopy, classification, deep learning, auto-encoder

1. Problem statement

Millions of star spectra are waiting for analysis in databases. There is a need for a universal model that will efficiently and effectively enable the vast majority of these spectra to be automatically reprocessed. The problem is not trivial as the appearance of the raw spectrum is influenced by many intertwined factors, including the resolution of the instrument, the pseudo-continuum shape, the presence of telluric lines, radial velocity, and last but not least, the physical parameters of the star. At the level of the observed spectrum, it is difficult to separate these factors. We believe that a well-designed auto-encoder and proper training will help distribute these effects in the latent space. The main idea is to use a deep neural network of the autoencoder type to extract in the latent space the factors influencing the appearance of the spectrum and then use them in downstream tasks: searching for similar spectra, clustering, classification or determining physical parameters.

2. Preliminary results

We have implemented a preliminary version of the model and performed training on 10,000 spectra from the UVES spectrograph. Then, for these 10,000 spectra and about 300 spectra from the UVES POP (Field stars, Bagnulo et al., 2003) project, we calculated the latent space vectors for which we applied the two-dimensional PCA (Pearson 1901) and t-SNE (van der Maaten and Hinton 2008). The points corresponding to the spectra from UVES POP are colored according to the spectral type. The PCA shows that the spectral type (aka effective temperature) is correlated with the principal components. In the case of t-SNE, we see the clustering of spectra into several distinct groups. The

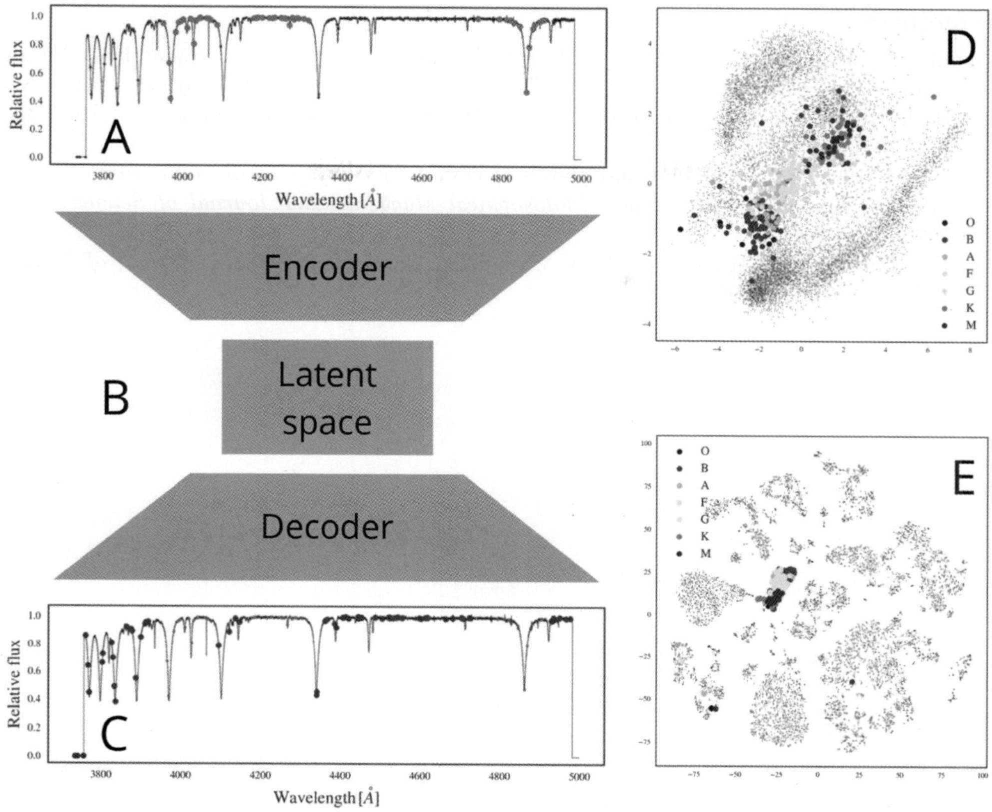

Figure 1. Panel A shows a stellar spectrum with samples (highlighted in red) that are fed into the auto-encoder neural network. **Panel B** shows the architecture of a deep neural network. It consists of a transformer based encoder and multilayer perceptron based encoder. The latent space symbolises a low dimensional representation of the input signal. **Panel C** shows (in blue) the samples where the model is making predictions. **Panel D** contains the principal components of the latent space with colour-coded spectral types taken from UVES-POP Field Stars dataset, the diagonal direction strongly correlates with the spectral type, and **Panel E** contains the results of t-SNE dimensionality reduction.

initial inspection at this stage has shown that the shape of the groups is determined by the instrumental effects. For the details see Figure 1. For this reason, our current work focuses on methods to better structure the latent space. Due to the complexity of the task, the shown results are promising.

Acknowledgements

This work was financed from budgetary funds for science in 2018–2022 as a research project under the program "Diamentowy Grant", no. DI2018 024648.

Supplementary material

To view supplementary material for this article, please visit https://doi.org/10.1017/S1743921323001254

References

Bagnulo, S., Jehin, E., Ledoux, C., Cabanac, R., Melo, C., Gilmozzi, R., & ESO Paranal Science Operations Team 2003, The UVES Paranal Observatory Project: A Library of High- Resolution Spectra of Stars across the Hertzsprung-Russell Diagram. *The Messenger*, 114, 10–14.

Pearson, K. 1901, LIII. On lines and planes of closest fit to systems of points in space. *The London, Edinburgh, and Dublin Philosophical Magazine and Journal of Science*, 2(11), 559–572.

van der Maaten, L. & Hinton, G. 2008, Visualizing Data using t-SNE. *Journal of Machine Learning Research*, 9(86), 2579–2605.

Machine Learning in Astronomy: Possibilities and Pitfalls
Proceedings IAU Symposium No. 368, 2025
J. McIver, A. Mahabal & C. Fluke, eds.
doi:10.1017/S1743921323000790

Is GMM effective in membership determination of Open Clusters?

S. N. Hasan[1], Md Mahmudunnobe[2], Priya Hasan[3] and Mudasir Raja[3]

[1]Department of Mathematics, Maulana Azad National Urdu University, Hyderabad 500032, India
email: hasan.najam@gmail.com

[2]Department of Physics, Minerva University, San Francisco, California 94103, USA

[3]Department of Physics, Maulana Azad National Urdu University, Hyderabad 500032, India
email: priya.hasan@gmail.com

Abstract. Determination of membership of star clusters is a very important criterion in their study as they effect determination of cluster parameters like radius, age, distance, mass functions, etc. In this paper, we apply the unsupervised method of Gaussian Mixture Modelling (GMM) to find membership of 9 open star clusters of varying ages and locations in the galaxy using Gaia DR3 data. We compare our results to help understand the efficiency of GMM. We find that this method works well with relaxed clusters with ages larger than their relaxation times they approximate Gaussians better.

Keywords. Machine Learning, GMM, Gaia EDR3, Star Clusters, Membership

---

1. Introduction

A new paradigm shift in open cluster membership determination is made possible by the unmatched accuracy of Gaia data, and many machine learning (ML) models, such as by Cantat-Gaudin & Anders (2020), are employed to achieve this. Unsupervised clustering algorithms do not have a quantifiable evaluation metric, which makes comparing different models difficult. Previous work on membership determination using GMM includes Agarwal et al. (2021); Gao et al. (2020); Mahmudunnobe et al. (2021). In this study, we define a measurable statistic called the Modified Silhouette Score (MSS) to assess how well unsupervised models do in determining membership. We examine the effectiveness of the Gaussian Mixture Model (GMM) to identify members from the Gaia DR3 data. We investigate the dependence of MSS on age, distance, extinction, galactic latitude and longitude, and other characteristics. We discover that GMM works best for clusters that are closer and relaxed.

MSS can be used to measure the performance of an unsupervised model, which outputs two sets of stars: members and field stars. For star cluster membership problem, we assumed a normal distribution (low SD) for members and an uniform distribution (high SD) for field stars in the feature space with K features.

$$MSS = \frac{1}{K}\sum_{i=1}^{K} \frac{SD_{member} - SD_{field}}{max(SD_{member}, SD_{field})}$$

MSS can range from -1 to 1. The higher the MSS value, the better the performance of the model to distinguish member (normal) and field star (uniform) group.

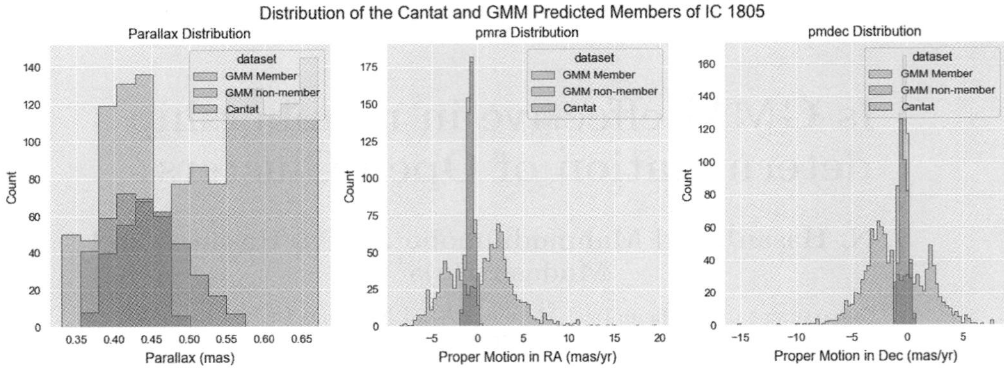

Figure 1. IC 1805: GMM distribution of field and cluster stars in parallax, proper motions (RA and Dec, respectively. A clear separation of the two components is visible.

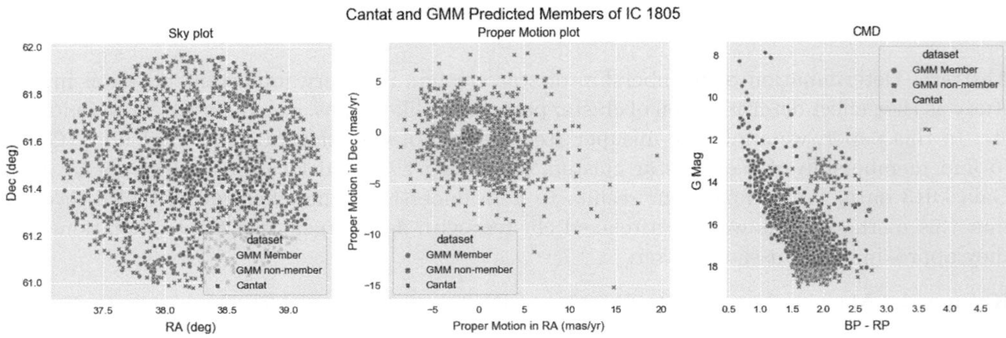

Figure 2. IC1805: shows the location of the new and old members in a skyplot, proper motion plot and a color magnitude diagram.

2. Analysis of IC 1805

Figures 1 and 2 show the analysis for IC 1805. A clear separation from member and field stars is obvious in the Fig. 1. Fig. 2 shows the location of the new and old members in a skyplot, proper motion plot and a color magnitude diagram.

3. Conclusions

The performance of the GMM method for membership determination can be evaluated using MSS. The performance of the model as assessed by and the cluster's distance were moderately correlated in this small sample, with GMM performing better for clusters that are closer to one another. There is little to no link between any other variables and GMM performance. In our complete paper, we provide a more thorough study utilising a bigger cluster sample†.

References

Cantat-Gaudin & Anders, 2020, *AA*, 633, A99

Agarwal et al., 2021., *MNRAS*, 502(2), 2582–2599

Gao. *ASS*, 365(2), 1–8, 2020.

Md Mahmudunnobe and Priya Hasan and Mudasir Raja and S. N. Hasan, 2021, *The European Physical Journal Special Topics*, 230, 10, 2177–2191

† The code used in this paper is available at `https://github.com/mahmud-nobe/GMM_membership`

Machine Learning in Astronomy: Possibilities and Pitfalls
Proceedings IAU Symposium No. 368, 2025
J. McIver, A. Mahabal & C. Fluke, eds.
doi:10.1017/S1743921323000923

Deep Radio Image Segmentation

Hattie Stewart[1,2], **Mark Birkinshaw**[1] **and Jason Yeung**[2]

[1]University of Bristol
email: hattie.stewart@bristol.ac.uk

[2]Rutherford Appleton Laboratory

Abstract. We show that the U-Net neural network architecture provides an efficient and effective way of locating sources in SKA Data Challenge datasets. The improved performance relative to PyBDSF is quantified and U-Net is proposed as an efficient source finder for real radio surveys.

Keywords. Radio Surveys, Machine Learning, Square Kilometre Array

1. Introduction

In the lead up to the operation of the Square Kilometre Array (SKA), the radio astronomy community has been trialing new methods of image processing and analysis suitable for handling the extremely large volumes of data expected from the new interferometer. The SKA Observatory (SKAO) has released model datasets in the form of simulations considered to be representative of future SKA data. In this work, we use the radio continuum images released in Science Data Challenge 1 (SDC1; Bonaldi *et al.* 2021). The objective of the challenge is to detect, classify, and measure the properties of all the sources within the images. Due to the large datasets, we apply Machine Learning (ML) techniques as a fast and optimised method of source extraction.

The task can be identified as an image segmentation problem. That is we wish to identify areas of interest within the images and classify them. For inspiration we look to the field of biomedical imaging, which hosts a wealth of image segmentation research. We apply U-Net, a neural network based on the encoder-decoder architecture, known for its ability to segment images quickly and accurately whilst only requiring a small dataset for training (Ronneberger, Fischer & Brox 2015). We present this promising method as a novel approach to source extraction in radio astronomical data.

2. Methods

In order to use U-Net for image segmentation we require a labelled training set. The SDC1 data consists of a set of 9 images taken at 3 different observing periods and frequencies. For each frequency, corresponding truth catalogues are provided which contain source information including location, shape and size information. To create our training set we take cutouts 256 x 256 pixels in size from all images of one observing frequency. We use the training catalogue provided, which includes source information for 5% of the field of view, to generate corresponding segmentation maps for each cutout. A segmentation map is an image of the same dimensions as the data cutout, with each pixel classified by a numerical system. In this case, we generate binary segmentation maps which use 0 to represent background and 1 to represent a source. We treat all sources as one class, thus are doing binary classification. We can create these segmentation maps by creating a zero array of dimensions 256 x 256 pixels. We then locate in the training catalogue

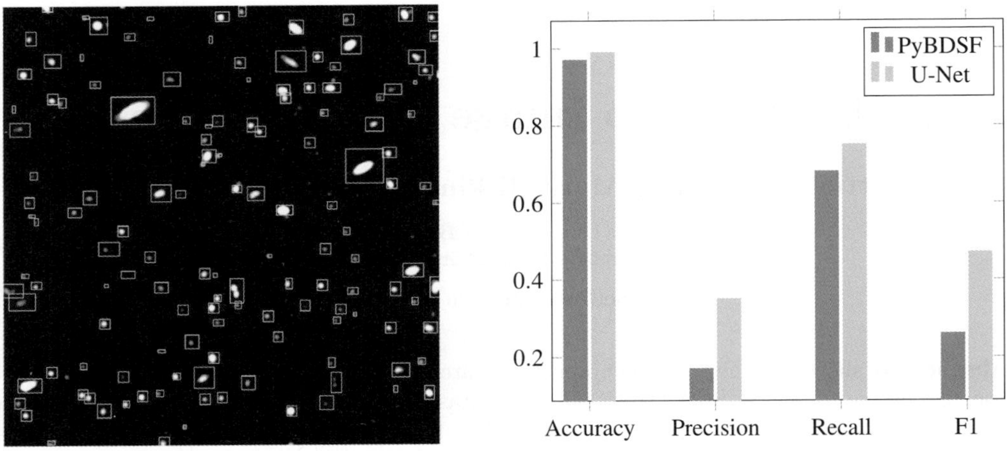

Figure 1. The left hand figure displays a cutout from the test dataset with the bounding boxes of the predicted sources overlaid. The right hand figure shows the comparative metrics of pixel classification for PyBDSF and U-Net. U-Net outperforms PyBDSF over all metrics. The metrics are as follows: accuracy is the proportion of correct predictions within all model predictions; precision is the proportion of the correct positive predictions in all cases classified as positive; recall is the proportion of the true sources the model has predicted; F1 Score is calculated from the precision and recall and varies between 0-1, it is proportional to both precision and recall. In all cases a higher score implies a better model fit.

all sources which exist within the data cutout, and update the zero array to 1 at these locations. We use both the binary segmentation map and the data cutout to train the U-Net, we refer to these as data-map pairs.

We split our data-map pairs 80:10:10 for training, validation and test. The training and validation set are used to train the network, whilst the test set is used for blind analysis of the model. The input data are reduced to a low dimensionality space within which source features are captured. The image is then reconstructed to the original input size. The binary cross entropy loss is minimised between the input and output segmentation. Thus the network learns to predict segmentation maps for new input data.

3. Results

We analyse the quality of the predicted segmentation by assessing the pixel classification accuracy. Since the data are heavily dominated by zero values (background), we also assess the source detection accuracy. We see that U-Net outperforms PyBDSF (Mohan & Rafferty 2015) for almost all metrics. We attribute the superiority of U-Net on pixel classification and source detection accuracy to its higher sensitivity to sources close to the background.

Initial results imply that U-Net is a highly promising method for source detection in radio images in the SKA era. However, careful tuning of the network parameters is required to ensure the true source distribution is recovered. In future work we will improve parameter estimation in the U-Net pipeline in order to recover the source parameters.

References

Bonaldi, A., An, T., Brüggen, M., *et al.* 2021, *MNRAS*, 500, 3821

Mohan, N. & Rafferty, D. 2015, *Python Blob Detection and Source Finder, Astrophysics Source Code Library*, ascl:1502.007

Ronneberger, O., Fischer, P. & Brox, T. 2015 in: N. Navab *et al.* (eds.), *Medical Image Computing and Computer-Assisted Intervention, Lecture Notes in Computer Science* 9351, p. 234 (Springer)

Machine Learning in Astronomy: Possibilities and Pitfalls
Proceedings IAU Symposium No. 368, 2025
J. McIver, A. Mahabal & C. Fluke, eds.
doi:10.1017/S174392132300073X

Computational techniques for high energy astrophysics and medical image processing

Nicolás Vásquez, Jennifer Ortega, David Erazo, Ricardo Caiza and Orlando Gutiérrez

Physics Department, Escuela Politécnica Nacional
email: nicolas.vasquez@epn.edu.ec

Abstract. We present two studies where astrophysics and computational techniques are applied to astroparticle physics and medical image processing. The first one is the use of software as Corsika, Gammapy, Astropy and ARTI to identify potential sites to study astroparticles physics in Ecuador. We determine the secondary particle fluence and M1 signal at different locations, and we calculated the effective area of a prototype Water Cherenkov Detector (WCD) in Quito. The second work is the manipulation of medical images using python and PyDICOM libraries.Aiming to explore the Computer Aided Detection (CAD) systems for the automated detection of numerous lung diseases. We develop a tool for extracting information from DICOM files more efficiently. A second tool consists of a lung segmentation technique for computed tomography images.These works shows the versatility of the computational tools encouraging young researchers to explore and further develop the use of such tools in other areas.

Keywords. Corsika, astroparticles physics, pulmonar segmentation

1. Introduction

Ecuador, a Latinamerican country located at zero latitude, has the advantage to study high energy astrophysics due to its geography. The study of high energy astrophysics involves sophisticated computational techniques which are natural for medical image processing techniques. Image processing and data analysis is now a basic tool for astronomers. To encourage students to explore alternative areas besides academics, we worked on two applications of the skills for data analysis: medical image manipulation and astroparticle physics.

2. Medical images and astroparticle physics applications

Image generation plays an important role in medicine. Since astronomy made use of images as source of information its data analysis techniques can be applied to other disciplines; we developed a DICOM visualizer and we explored morphological operation on images from TC scans, see figure 1 and we studied the visibility of a gamma source.

A TC scan collects information of radiation absorption and interaction with tissues in Hounsfield units and produces images. The segmentation process uses the pixel matrix in Hounselds units to initially identify lung tissue through its degree of linear X-ray attenuation. Then, we apply the marker-based Watershed algorithm, which is the main basis of the technique. In addition, different morphological operations and transformations are used on the axial sections for obtaining the lung image. Our visualizer is based on python and pydicom libraries, and a segmentation tool for lungs was created based on Watershed transform (Shojaii et al. (2005)).

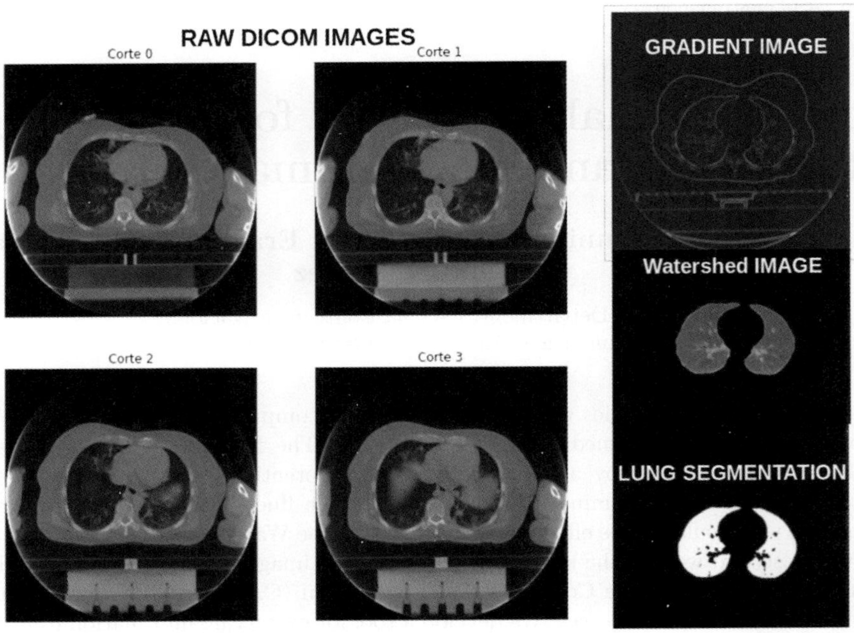

Figure 1. Lung segmentation procedure: We applied two morphological operations to raw DICOM files.

We simulated the cosmic ray background and the gamma emission from Crab at 4100 a.s.l in Quito for 6 hours, which is the transit time of M1 over the equatorial sky. For this simulations we used ARTI (Sarmiento-Cano et al. (2020)) to determine the cosmic ray background with its chemical composition, gammapy (Nigro et al. (2019)) to determine the photon flux in the 1 to 10 TeV range, Corsika (Heck et al. (1998)) to calculate the secondary particle fluence from both sources, and Astropy (Astropy Collaboration et al. (2018)) to determine the transit of Crab pulsar(M1). We found a excess in muons and gamma photons from M1 that would allow the indirect detection of the source at ground level.

3. Conclusions

The DICOM visualizer opens files without installing a external software and the segmentation tool works well. We reported that the segmentation tool can identify abnormal tissue in the lung (black structures) that could help to diagnose illnesses an open the application in an automated diagnosis tool. This work is a summary of 4 graduation projects of undergraduate students of the physics department of the Escuela Politécnica Nacional, for further information visit the following links.
`https://github.com/jennifer95/visualizacion-segmentacion-pulmonar-DICOM-python`
`http://201.159.223.36/poli/index.php`

References

Heck D., Knapp J., Capdevielle J. N., Schatz G., Thouw T., 1998, cmcc.book

Sarmiento-Cano C., Suárez-Durán M., Calderón-Ardila R., Vásquez-Ramírez A., Jaimes-Motta A., Núñez L. A., Dasso S., et al., 2020, arXiv, arXiv:2010.14591

Nigro C., Deil C., Zanin R., Hassan T., King J., Ruiz J. E., Saha L., et al., 2019, A&A, 625, A10. doi:10.1051/0004-6361/201834938

Astropy Collaboration, Price-Whelan A. M., Sipőcz B. M., Günther H. M., Lim P. L., Crawford S. M., Conseil S., et al., 2018, AJ, 156, 123. doi:10.3847/1538-3881/aabc4f

R. Shojaii, J. Alirezaie and P. Babyn, IEEE International Conference on Image Processing 2005, pp. II-1270, doi: 10.1109/ICIP.2005.1530294.

Machine Learning in Astronomy: Possibilities and Pitfalls
Proceedings IAU Symposium No. 368, 2025
J. McIver, A. Mahabal & C. Fluke, eds.
doi:10.1017/S1743921323001060

Deep learning proves to be an effective tool for detecting previously undiscovered exoplanets in Kepler data

Amelia M. Yu

Henry M. Gunn High School, PAUSD, Palo Alto, CA 94306, USA
email: amelia.yu7@gmail.com

Abstract. Using deep learning (DL) with the Adam optimizer, I detected 15 previously undiscovered exoplanets in the Kepler data collected by the National Aeronautics and Space Administration (NASA). Some of the exoplanet transit signals were evident, but others were not as strong. Further evaluation is needed. By using my own code and DL libraries including TensorFlow, I built a Python program to search for exoplanets. Among the new candidate exoplanets detected by my program, 13 of them are ultra-short-period (USP) exoplanets with orbital periods shorter than a day. Moreover, I experimented on this Python DL program with current candidate and confirmed exoplanets in the NASA database and was able to detect more than 200 candidate exoplanets and 94 of the 116 confirmed exoplanets in the NASA database. These findings show that DL can be an effective tool to detect objects of interest, such as exoplanets, in astronomy big data.

1. Introduction

The NASA Kepler space telescope observed 530,506 stars for 9.6 years in order to find exoplanets and collected a huge amount of data but posed a question for us: How can we efficiently detect exoplanets in this big data? There have been efforts to apply artificial intelligence, and more specifically, DL, in the search of exoplanets. Pearson et al. (2017) reported that their convolutional neural network (CNN), a type of deep learning (DL), was capable of detecting Earth-like exoplanets in noisy time-series data with a higher accuracy than a least-squares method. Zucker et al. (2018) examined the feasibility of using CNNs to detect planetary transits in light curves plagued by red noise. I also used DL to search for exoplanets. However, my method has yet to be found in literature.

2. Research Methodology

My data was retrieved from the public NASA Kepler database. Then, I used the transit method and implemented DL, particularly a CNN, to search for exoplanets in the data. By using my own code, DL packages, and open source libraries such as TensorFlow, I developed a Python program. The program first normalizes the light curves, folds these light curves to intensify the transit signals, and meanwhile trains the DL model using the Adam optimizer. For folding, I initially folded the normalized light curves 100,000 times, with each period spaced evenly between a set range of days in order to overlap the possible periodic transit signals in the folds. For example, by initially producing 100,000 phase folds in a time range of 0–10 days, there would be a $10 \div 100{,}000 = 0.0001$ day difference between fold n1 and fold n2 in a time series. After the first detection of a possible transit, either the total fold number would be adjusted to become larger, the

Table 1. Fifteen newly detected exoplanets (Stellar data from the Open Exoplanet Catalog and the NASA Exoplanet Archive.)

KIC	Orbital Period (days)	Transit Score	Distance (parsec)	Stellar Effective Temperature (K)	Stellar Radius (Solar radii)	Stellar Mass (Solar mass)
5881688	0.838103681	1.0	1475 ± 60	$5903 + 107 - 160$	$1.08 + 0.29 - 0.12$	$1.06 + 0.08 - 0.07$
10975146	2.332733327	0.999	343 ± 15	$4653 + 51 - 57$	$0.76 + 0.03 - 0.05$	$0.71 + 0.02 - 0.04$
4770365	0.8046611467	1.0	$543.444 + 10.777 - 10.373$	4303.0 ± 122.0	0.714145	0.67
8780959	0.8724768248	0.9999999	3025 ± 41	$6846 + 320 - 290$	$1.54 + 0.21 - 0.18$	1.9 +0.6 -0.4
6871071	0.9489352894	0.9999976	951	5977 ± 200	0.9 ± 0.5	1.03 ± 0.11
10024051	4.0416704167	1.0	523 ± 6	$5063 + 95 - 81$	$0.74 + 0.05 - 0.07$	$0.78 + 0.05 - 0.06$
7871954	0.968497885	1.0	192.3	$3944 \pm 150 - 50$	0.48 ± 0.03	0.575 ± 0.016
9595827	0.9762298623	1.0	940 ± 20	5543 ± 138	0.89 ± 0.07	0.92 ± 0.08
2692377	0.7708622086	1.0	437	5539 ± 73	1.11 ± 0.12	0.99 ± 0.06
3338885	0.9852488525	1.0	431 ± 4	$4897 + 91 - 107$	0.75 ± 0.03	0.80 ± 0.04
8565266	0.9161362614	0.9995336	973 ± 18	$5820 + 227 - 188$	$1.01 + 0.20 - 0.11$	$1.00 + 0.09 - 0.07$
9823519	0.8833966340	1.0	1122 ± 19	$3827 + 45 - 60$	$0.53 + 0.03 - 0.05$	$0.55 + 0.03 - 0.05$
6922244	0.8806246062	1.0	1330 ± 180	6251 ± 75	$1.45 + 0.12 - 0.13$	$1.13 + 0.09 - 0.10$
7825899	0.9011871119	0.9999984	851	5175.0 ± 25.0	0.768535	0.88
12470954	0.6340923409	1.0	704 ± 9	$5447 + 76 - 67$	$0.92 + 0.06 - 0.04$	$0.93 + 0.04 - 0.03$

Table 2. Sample of the results of my program for confirmed and candidate exoplanets.

	The results of my program		NASA data	
KIC	**Transit signals**	**Period (days)**	**Period (days)**	**Status**
1725016	Detected	7.4073440734	7.40742502	Confirmed (Kepler-748 b)
6964929	Detected	0.6650301503	0.665025941	Confirmed (Kepler-1340 b)
5942808	Detected	0.63	0.63	Candidate
4665571	Detected	0.77	0.77	Candidate

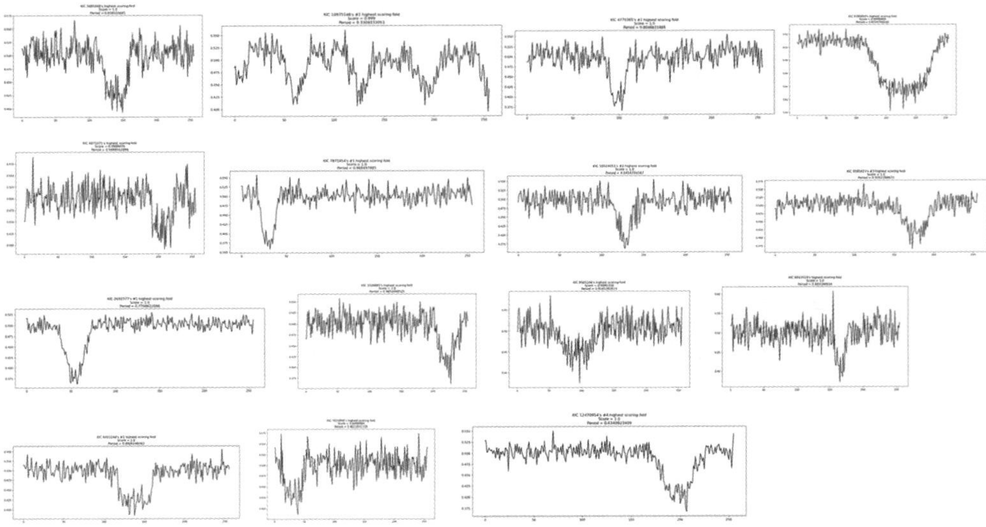

Figure 1. Transit signals detected in this program for new, previously undetected exoplanets orbiting KIC 5881688, KIC 10975146, KIC 4770365, KIC 8780959, KIC 6871071, KIC 10024051, KIC 7871954, KIC 9595827, KIC 2692377, KIC 3338885, KIC 8565266, KIC 9823519, KIC 6922244, KIC 7825899, and KIC 12470954 (X-axis: time; Y-axis: flux.)

time range for the folds would be adjusted to a shorter range around the possible period, or both the fold number and time range would be adjusted to generate a smaller time difference between folds for more time-wise precision. I then used this DL program to search light curve folds for exoplanet transits.

3. Results - Detection of 15 New and Several Hundred Confirmed or Candidate Exoplanets

This program detected more than 200 candidate exoplanets and 94 of the 116 confirmed exoplanets in the NASA database, with a 100% detection rate for all confirmed USP exoplanets and an 81.03% detection rate for overall confirmed exoplanets listed in the NASA database, proving its validity and effectiveness when tested on Kepler data. Furthermore, my program detected 15 new, previously undiscovered exoplanets for further evaluation. Tables 1 and 2, as well as Figures 1 and 2, show the results of detection for these 15 new candidate exoplanets, a sample of the confirmed and the known candidate exoplanets, respectively, as indicated in the captions of the tables and figures.

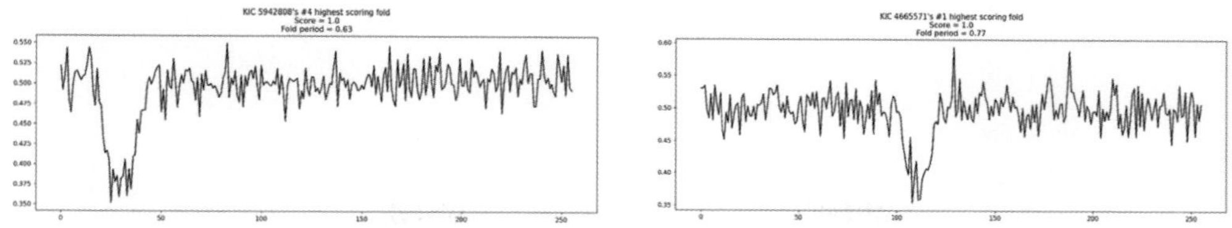

Figure 2. Transit signals detected in this program for candidate exoplanets orbiting KIC 5942808 and KIC 4665571.

4. Discussion and Conclusion

Besides the 15 new candidate exoplanets, my program could also detect more than 300 hundreds of the previously discovered exoplanets in the Kepler data, demonstrating its capability to detect real exoplanets. To conclude, the detection of the known exoplanets and the new candidate exoplanets in this research demonstrates that deep learning is a feasible method to discover exoplanets in astronomy big data, such as the Kepler data.

References

Pearson, K. A., Palafox, L., & Griffith, C. A. (2017). Searching for exoplanets using artificial intelligence. Monthly Notices of the Royal Astronomical Society, 474(1), 478–491. https://doi.org/10.1093/mnras/stx2761

Zucker, S., & Giryes, R. (2018). Shallow transits—deep learning. I. feasibility study of deep learning to detect periodic transits of exoplanets. The Astronomical Journal, 155(4), 147. https://doi.org/10.3847/1538-3881/aaae05

(Due to the page limit of this poster publication, many details are not reported here. For more information on this research, please refer to my contributed talk presentation/publication in Focus Meeting 8 of the IAU General Assembly XXXI, 2022.)

Author index